- 3

The Team Developer:
An Assessment and
Skill Building Program

Student Guidebook

Jack McGourty
Columbia University

Kenneth P. De Meuse
University of Wisconsin – Eau Claire

INTRODUCTION

The Team Developer is an electronic assessment and feedback system designed to help you grow and develop as a team member and student. It was developed nearly ten years ago, and since that time has been used in both academic and industrial settings. Individuals who have used the assessment package uniformly tell us that it has been a worthwhile experience. The format is designed to provide both the giver and receiver of feedback a safe, nurturing learning environment. *The Team Developer* enables all team members to share concerns, issues, and evaluations in a constructive fashion. When used properly, this feedback process will enhance team member communication and improve team performance.

The use of teams has become increasingly commonplace in undergraduate and graduate classrooms during the past decade. Nowadays, it seems that courses in every discipline, ranging from engineering and business to nursing and humanities, have a team component as part of the curriculum. Instructors are trying to help students develop the skills required to be successful in today's team-oriented workplace. Unfortunately, little class attention is given to the skills needed to perform well in teams. Students simply are given a "team assignment" and expected to automatically work together effectively. Little or no class time and course resources are devoted to such issues as: (a) How do we get a group of individuals with different motives, likes, dislikes, personalities, and needs to work together? (b) What should you do when team members do not perform the work they promised? (c) How do we conduct group meetings? (d) How do you get someone to stop dominating our team? (e) How do you get someone to contribute his or her thoughts and ideas? (f) How are we getting evaluated? The list goes on and on! Such a sink-or-swim mentality rarely yields the desired results. Indeed, some of a student's worst nightmares about college come from working on so-called "team projects."

In many cases, instructors themselves are not fully trained in team concepts. Their expertise appropriately lies in their technical specialty. Consequently, students receive neither the instruction nor feedback required to learn team skills.

In this Guidebook, we address two fundamental needs of students placed into classroom teams. First, we review some of the key skills needed to be an effective team member. In the chapters ahead, we present material on the principles of highly successful teams, the stages of team development, how to conduct team meetings, how to be a good listener, how to resolve group conflict, and so on. Secondly, this Guidebook and the accompanying software package provide students an opportunity to give (and receive) feedback from team members in a safe, constructive setting. This feedback will help the student and team become more successful in future projects. Our overall goal is to provide students with a reference manual that will be a foundation for their continued growth and development.

This book is intended for the student who works on teams to accomplish a task or assignment. Both undergraduate and graduate students will benefit from the concepts reviewed. *The Team Developer* assessment and feedback program will help students recognize areas where they are somewhat deficient and offer them constructive suggestions on how to develop. This book is intended as a supplemental text. It should serve as a handy reference for team activity.

We have designed it to serve as a guide for team-related projects in any course regardless of discipline. Therefore, we hope students will refer to it often, whenever taking courses requiring a team component.

We want to thank many individuals for their insights, contributions, and encouragement in creating *The Team Developer* software and writing this book including Eli Fromm, the Principal Investigator of the Gateway Engineering Education Coalition (Drexel University), Richard R. Reilly (Stevens Institute of Technology), and Peter Dominick (Columbia University). Michael Kelly (Yale University), Margaret Kelly and Jenny Lee (both from Columbia University) were instrumental in ensuring that this book was properly edited and formatted. Finally, we are very grateful to the hundreds of individuals who have taken *The Team Developer*, and contributed to its construction and refinement during the past several years.

This work has been sponsored by the Gateway Engineering Education Coalition and supported, in part, by the Education and Centers Division of the National Science Foundation (Awards EEC-9109794 and EEC-9727413).

 Gateway Engineering Education Coalition

CHAPTER 1

THE IMPORTANCE OF TEAMS AND TEAMWORK IN THE 21st CENTURY

Corporate America is having a love affair with teams.
And why not? When teams work there's nothing like them
for turbocharging productivity.

Brian Dumaine
Fortune Magazine

It has been estimated that nearly all of the Fortune 500 companies use teams of some form (e.g., project teams, self-directed work teams, problem solving teams, quality circle groups) in their business (Dumaine, 1994; Lawler & Cohen, 1992). Xerox, for example, reportedly has more than 7,000 quality-improvement teams involving about 75% of their employees. Florida Power and Light Company reportedly has 1,900 quality teams, and nearly every employee participates on a team (Harrington-Mackin, 1994). In today's corporate environment, it appears that the team (not the individual) holds the key to organizational success.

Teams add a powerful dimension to the workplace. Teamwork combines the skills and the creativity of a diverse number of people in order to produce an effective outcome. At virtually every level of the organization, from manufacturing on the shop floor to decision making in executive boardrooms, teams (not individuals) are doing the work. Consequently, more and more employers are looking for new recruits who possess not only the requisite technical skills but also the ability to work well in the self-directed, collaborative environment inherent in teaming.

There are a number reasons why teams have become prevalent in today's workplace. First, the pressure on businesses to respond to increased global competition has stimulated a search for new ways to work more efficiently and effectively. A prominent aspect of effectiveness is meeting customer needs. Proponents of managerial initiatives such as total quality management (TQM) and process re-engineering recommend teams as the preferred way to organize and accomplish work (Lawler, 1994; Hammer & Champy, 1993). Similarly, concurrent engineering approaches rely on cross-functional teams of researchers to enhance product development and innovation (McGourty, Tarshis, & Dominick, 1996).

Competitive pressures also have led to wholesale organizational changes such as corporate restructuring and downsizing. Because smaller, flatter organizations require employees to be more flexible and to play a greater role in deciding how work gets done, self-directed work teams have become increasingly popular (Manz & Sims, 1993; Wellins, Byham, & Wilson, 1991). Finally, many jobs and projects have become increasingly complex, making it difficult for one person to perform them. Hence, many jobs today require the use of teams as the basic work unit (e.g., airplane crews, surgical units, research and development teams).

In summary, teams have become a vital part of the contemporary workplace, because they enable companies to best meet the demands of their markets. Teams allow corporations to improve customer satisfaction, enhance products and services, and increase productivity. Team-based organizations are leaner, more efficient, and can respond more rapidly to change than traditionally structured companies.

A Team versus a Collection of Individuals

Fifty-thousand screaming fans at a football game, hundreds of shoppers searching the sales racks at Macy's for the best bargains, literally thousands of drivers sitting in their cars on the Dan Ryan Expressway during rush hour — all are good examples of a collection of people *but not a team*. Perhaps, we could add to this list the student group you were assigned to work with in your engineering class. *In other words, simply putting together a collection of people in one space and time does not denote a highly effective, interactive, cooperative, problem-solving, conflict-free work team.*

A work team is a group of individuals whose work is interdependent and who are collectively responsible for accomplishing a performance outcome (Mohrman & Mohrman, 1997). This outcome might be (a) delivering services, (b) designing, developing, and producing products, (c) implementing improvements and innovations, or (d) analyzing and writing up a case study for your management professor. Teams come in all shapes, sizes, and compositions. Regardless of the work performed, all teams are characterized by the following features:

- **A dynamic exchange of information and resources among team members**
- **Task activities coordinated among individuals in the group**
- **A high level of interdependence among team members**
- **Ongoing adjustments to both the team and individual task demands; and**
- **A shared authority and mutual accountability for performance (i.e., results).**

Years ago, jobs were structured around functional duties and responsibilities. Employees were assigned to a specific department, such as accounting, engineering, manufacturing, or marketing. Individuals were committed to their particular piece of work and unlikely to see themselves as sharing responsibility for all aspects of a project, product, or service. Managers spoke in terms of "individual contributors" and "functional units." During that era, it made sense to focus only on the specific activity surrounding your job. As a student, you needed to develop only those skills necessary for that one position.

A new paradigm is emerging in the workplace of the 21st Century (Sundstom, De Meuse, & Futrell, 1990). In this paradigm, employees must be sensitive to the demands of the entire business. The so-called "silo mentality" of yesterday where an employee was concerned with only his or her unique job assignment is vanishing. There is a new ballgame being played on the sandlots of Corporate America, and it is important to recognize that simply because you were successful in the old paradigm, does not automatically make you effective in the new one. Just ask Michael Jordan! Here is perhaps the greatest basketball player of all time, who had many, many difficulties adjusting to the demands of baseball (eventually giving up and returning to basketball).

Figure 1.1 It's a whole new ballgame – Workplace 2000

The Success and Failure of Teams

The glowing successes of work teams have been widely publicized in the media. Stories trumpet such achievements as follows:

- Federal Express and IDS boost productivity up to 40% by adopting self-managed work teams (Dumaine, 1994).
- AT&T's operator team increases service quality by 12% (Wesner & Egan, 1991).
- Dana Corporation's valve plant reduces customer lead-time from six months to six weeks by using teams (Sheridan, 1990).
- Proctor and Gamble lowers manufacturing costs 30-50% by the use of self-directed work teams (Fisher, 1993).
- Teams increase output 280% at Honeywell (Fisher, 1993).
- Publication teams at GTE increase production of telephone directories 158%, while decreasing errors by 48% (Joinson, 1999).

Despite the purported success of the team approach, there are some researchers who claim teams can be problematic to manage and may often lead to poor results (Bergmann & De Meuse, 1996; Dumaine, 1994; Shaw & Schneier, 1995). In a *Fortune Magazine* article entitled "The trouble with teams," Brian Dumaine asserted that although in theory everybody loves them, in practice it's another story (1994, p. 86). According to one scientific study, nine out of ten teams fail (*The Team-Based Organization*, 1994).

Overall, we can conclude two things. First, teams are extremely popular in Corporate America today. Second, although there might be some mixed reviews regarding how successful teams are in companies, work teams will likely remain very pervasive in organizations during the coming decade. Consequently, students would be prudent to develop team-relevant skills as they prepare for their careers.

Some research suggests a key reason why teams fail is that employees are ill-prepared to make the transition from individual contributor to team member. Thomas Bergmann and Kenneth De Meuse (1996) investigated the attempted implementation of self-managed work teams in a food processing plant. They discovered that employees lacked the fundamental team skills of problem solving, dealing with conflict, conducting effective meetings, and interpersonal communication. The employees resisted the movement to self-managed work teams to such an extent that after 10 months management returned to the old system of production.

Obviously, being an effective team member requires a unique set of skills, ones that people do not necessarily come by naturally. However, like most skill sets, *team* skills can be learned and practiced. Keys to learning these skills include (a) proper instruction on core concepts and principles, (b) opportunities to practice and apply these concepts, and (c) access to feedback on one's performance. In this Guidebook, we hope to provide you with the tools and feedback to become an effective team member. Once accomplished, you can use this experience to better market yourself to college recruiters who are demanding these skills for various job openings.

CHAPTER 2
UNDERSTANDING TEAM PROCESSES, ROLES, AND BEHAVIORS

And the Lord said to the rabbi, "Come, I will show you hell." They entered a room where a group of people sat around a huge pot of stew. Everyone was famished and desperate. Each held a spoon that reached the pot, but each spoon had a handle so long that it could not be used to reach each person's mouth. The suffering was terrible!

"Come, now I will show you heaven," the Lord said after a while. They entered another room, identical to the first — the pot of stew, the group of people, the same long spoons. But there everyone was happy and nourished.

"I don't understand," said the rabbi. "Why are they happy here when they were miserable in the other room and everything was the same?"

The Lord smiled. "Ah, but don't you see?" He said. "Here, they have learned to feed each other."

> *Merle Shain*
> *Author*

The Green Bay Packers, the Chicago Bulls, the Atlanta Braves, the New York Yankees, the San Francisco 49ers — all sports teams with an established record of winning during the 1990s. What was the key to their success? What characteristics did these teams possess that losing teams did not? Is there some magic formula or equation that we can use to develop successful work or student teams? How have *they* learned to feed each other?

In this chapter, we will examine the processes, roles, and behaviors that underlie team performance. We will identify some of the fundamental principles of highly effective teams. We also will explore various stages teams go through to become effective, recognizing that team success does not happen overnight but requires an ongoing process of growth and development. In the next chapter, we present a model of team performance and examine four key behaviors (skill sets) in this model.

Principles of Highly Effective Teams

One way of vividly seeing what a highly performing team looks like is to examine a dysfunctional one. In such a case, you will first of all find little or no commitment to the team's overall purpose or mission among the team members. They spin their wheels and get very little accomplished. They think and act like a group of individuals rather than as a team. They fail to challenge the status quo, and they do not view themselves as responsible for following through on agreed upon actions. In many instances, team members lack the structure and direction required to succeed. Their goals are not clearly defined, for example, or they have no milestones and no real action plans. Meetings lack agendas and structure; confusion reigns over roles and expectations.

In addition to these structural problems, members of poorly performing teams also lack an understanding of how their individual behavior can impact the team's overall success or failure. Frequently, members are much more interested in satisfying their personal (selfish) needs than meeting team goals. Consequently, even though team members may have excellent technical skills, they lack the team skills necessary for transforming their technical competence into team success. Instead, individuals compete to be heard or just one person dominates the discussion. In either case, conflict resolution is poor and team synergy is nonexistent.

Think of what happened in a student group you were a part of last semester. Agreed upon meeting times when people failed to show (or were late). Students not doing the work they promised so you had to do it yourself. Two-hour meetings where nothing appeared to be accomplished. Assignments handed in late because team members didn't do what was promised. Confusion with regard to who was doing what and when. "Big-mouthed" students telling everyone what to do and voicing their opinion on every topic, while some students do not say a word. Do these characteristics ring a bell?

Carl Larson and Frank LaFasto investigated dozens of teams during a three-year period. Their sole focus was to ascertain what were the characteristics of effectively functioning work teams. During the course of the study, they interviewed members and leaders of an extraordinarily diverse number of teams, ranging from cardiac surgery teams, executive management and project teams, airline cockpit crews, mountain climbing teams, and engineering teams. They wrote the results in a book entitled, *Teamwork: What Must Go Right / What Can Go Wrong* (1989). Overall, Larson and LaFasto found successful teams have the following eight critical factors:

1. **A clear, challenging goal; this goal gave the group members something to shoot for. The goal was *understood* and *accepted* by the entire group.**
2. **A results-driven structure (i.e., members had clear roles and accountabilities and there was an effective communication system within the team).**
3. **Competent, talented team members.**
4. **Unified commitment; in other words, members put the *team* goals ahead of *individual* needs.**
5. **A positive team culture. This factor consisted of four elements: (a) honesty, (b) openness, (c) respect, and (d) consistency of performance.**
6. **Standards of excellence (i.e., an expectation of high performance, to be successful as a team).**
7. **External support and recognition. Effective teams receive the necessary resources and encouragement from parties outside of the group.**
8. **Effective leadership. Simply put, successful teams have good leaders.**

Of these eight characteristics, the authors reported that the most important one was that teams have a well-defined, specific goal for which to strive. Unless groups have an over-arching team goal (or set of goals), confusion with regard to task accomplishment is likely to result. As students, you can hold all eight of these factors as guiding principles to achieve in your team.

Stages of Team Development

One key to developing a highly performing team is to remember that successful teams — whether they are airline crews, R & D units, or student groups — do not occur automatically or overnight.

They take a lot of effort and time! You need to be aware of how groups form and evolve. You should recognize that all teams go through growing pains. In fact, many teams never reach full maturity. Instead, teams often flounder in a morass of personal agendas, misdirection, poor leadership, and little synergy. You also must realize that teams are composed of *individuals* that not only look different but are different. Obviously, team members differ in terms of physical characteristics such as gender, race, age, height and weight. However, members also differ in terms of personal needs, hot buttons and cold buttons, personalities, leadership preferences, motives, likes and dislikes, how they cope with stress and so on. Individual differences that you can see are not nearly as important as those differences you cannot see.

It is generally acknowledged that there are five stages or phases of team development (Tuckman, 1965; Tuckman & Jensen, 1977). These five stages are predictable and fairly observable to knowledgeable team members. As a team evolves from a collection of individuals (Stage 1) into a highly effective team (Stage 4), its maturation is similar in many respects to an infant growing into an adult.

Figure 2.1 The challenge of getting a collection of diverse individuals to effectively work together as a team

A wise leader, just like a wise parent, recognizes that the needs of a group, or child, change over time and alter guidance accordingly. Likewise, if you as a team member know what to expect, you can better understand what is going on, why it is happening, and act accordingly.

Stage 1 — Forming

The formative stage involves the transformation of individuals into team members. Team members naturally struggle at this stage with (a) defining the nature of the task to be completed, (b) knowing each other's personalities, motives, work styles, etc., (c) determining acceptable and unacceptable group behavior, (d) deciding what information and resources are necessary to perform the given objective, and (e) simply trying to determine where, when, and how to begin. In most instances, conversation among team members is polite but guarded. The group tends to be quite passive and dependent on a leader (the course instructor) to tell them what to do. As in infancy, when our parents needed to provide the structure and proper nurturing for us to grow, a team in Stage 1 is completely dependent on the course instructor to tell them what to do and help show members how to accomplish it. Unfortunately, course instructors frequently fail to provide the proper setting for student teams to start off on the right foot.

Stage 2 — Storming

As the team progresses, it goes into a stage similar to adolescence. In this phase, team members vie for control over the group much like teenagers attempt to control their own life. Members begin to think they know more than they do, set unrealistic expectations regarding their performance, and typically argue openly amongst themselves. Oftentimes, the group has experienced some success and now believes it is ready "to take on the world." It is an emotionally charged time. On one hand, there is high energy, enthusiasm, and optimism. On the other hand, there can be anger, resentfulness, and restlessness. Polarization and scape-goating occur as the team tries to identify itself. Frequently, chastised members will voluntarily (or be forced to) leave the group. Stage 2 is a very challenging time for the team. However, it is a necessary condition for the team as it matures to the next stage of development.

Stage 3 — Norming

During this stage, team members will have resolved their basic differences and begun to work together. Some members may have left the group. The remaining ones are far more accepting of each other, their own roles, and the team's goals and objectives. A spirit of friendliness, cooperation, and mutual respect characterizes this stage. Members share information willingly, communicate openly, and solve problems effectively. Individuals begin to understand the different strengths and weaknesses of various team members and consider them in task assignment. Members start to identify as a group (i.e., become cohesive). They often socialize and begin to become close friends.

Stage 4 — Performing

Team members feel an intimacy with each other during this phase and gain a great deal of satisfaction from the exchanges that are possible within the team. There is no issue over power, control, or status. When problems occur, members accept them as part of the interpersonal dynamics of the team process and openly work through them. Members are comfortable with their roles and with each other; they work as a unit to accomplish assignments. The team is productive, efficient, and highly focused. Credit for success is shared. The members recognize a synergy occurs when they work together.

Stage 5 — Adjourning

The final stage in team development is "adjourning." This phase may or may not occur during a team's life cycle. In many instances, a team will accomplish its mission and disband. This period is a time of letting go and moving on with our individual lives. Members recognize that their time as a group is over. No matter how committed the individuals are to keeping in contact, their journey as a work team has ended. Most members will experience a sense of loss. Consequently, there is much sadness and grieving. Yet, it also is a very happy occasion. The team is proud of what they attained and members know they were part of a very special group. As a whole, this stage is characterized as a time of mixed emotions. If at the end of the semester, you experience no remorse that your team is breaking up, it is doubtful that you ever reached Stage 4.

STAGES OF TEAM DEVELOPMENT

FORMING

STORMING

NORMING

PERFORMING

ADJOURNING

Figure 2.2

The five stages of team development are visually depicted in Figure 2.2. The faces are designed to reflect the different phases of emotion a team goes through as it evolves and matures. Hopefully, they will enhance your understanding of the interpersonal dynamics in teams. Overall, you can anticipate two outcomes in the team to which you have been assigned. First, the team will experience various stages of activities and emotions as the semester unfolds. Your awareness of these stages as a natural evolution of team maturity will prepare you to better perform in that environment. Second, a team gradually performs at a higher level as it matures (see Figure 2.3). Having this knowledge should enable you to guide your group through the beginning stages of group formation more quickly and effectively. Your behavior can make a huge difference in leading the team. In the next chapter, we will discuss four key behaviors that will assist you in this effort.

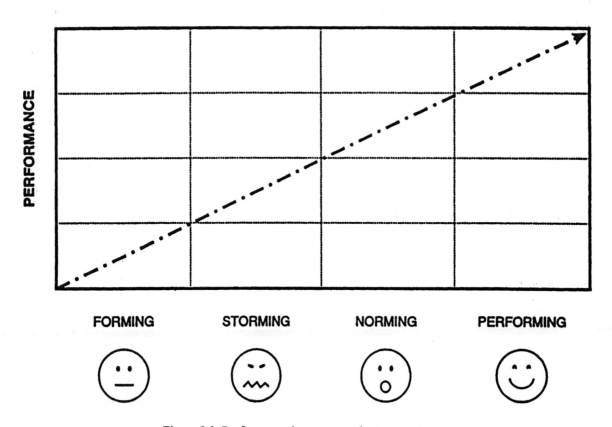

Figure 2.3 Performance increases as the team matures

CHAPTER 3
A MODEL OF TEAM PERFORMANCE

A single twig breaks,
but a bundle of twigs is strong.

Tecumseh [1768-1813]
Native American Chief

A team's ability to perform effectively depends on many, many factors. As mentioned in the previous chapter, the stage of its development is one factor that affects how well a team performs. Other *internal* factors such as team cohesiveness, how well members communicate, and team leadership also influence how successful teams work. On the other hand, some factors impacting team performance may be *external* to a team and, therefore, beyond the direct control of its members. For example, an organization's culture can significantly support or oppose teamwork. It can be very difficult for team members to develop trust in one another if they work in a company where employees are reprimanded for speaking openly.

In this chapter, a model of team performance is presented. This model highlights both *internal* and *external* factors and their effect on team performance. Four specific behaviors are identified with regard to their impact on performance: (a) communication, (b) decision making, (c) collaboration, and (d) self-management. Special attention is given to defining those behaviors and describing various roles members need to maximize team performance.

Team Performance Model

Figure 3.1 depicts a concentric circle model of team performance. It illustrates at a glance how several major factors work in concert to influence team performance. In the outer ring, various factors *external* to the work team are identified. The next ring presents four specific behaviors that members can perform in teams; these behaviors are *internal* or under the direct control of a team's members. Various roles team members should play as they perform these four behaviors are illustrated as well. Finally, the innermost circle represents the overall performance level of a given team. Team performance encompasses such achievements as quality and quantity of work, innovation, meeting imposed project deadlines, and so on.

As you can see, the outer circle in Figure 3.1 consists of the following five variables:

- **Organizational Structure;**
- **Rewards;**

- **Task Assignment;**
- **Resources; and**
- **Corporate Culture.**

Organizational Structure represents the formal pattern of how employees and jobs are grouped. Some organizations are very tall hierarchies with many tiers of management; others are relatively flat with few layers of supervision. An "organizational chart" often illustrates the structure of a company. Rewards represent the manner in which an organization reinforces its employees' good performance (e.g., merit pay increases, bonuses, promotions, etc.). *Task Assignment* refers to how a company allocates specific tasks, duties and responsibilities to its employees. Some companies carefully segment work into incremental steps and assign small tasks for each employee to perform (e.g., assembly-line jobs). Companies will empower employees to perform entire series of jobs or may require the work to be performed by a group of employees. *Resources* simply refers to the amount of financial assets an organization possesses. *Corporate Culture* can be defined as the system of values, beliefs, and norms that exists in any organization. An organization's culture represents the informal, philosophical side of a company, which can largely affect how it conducts business.

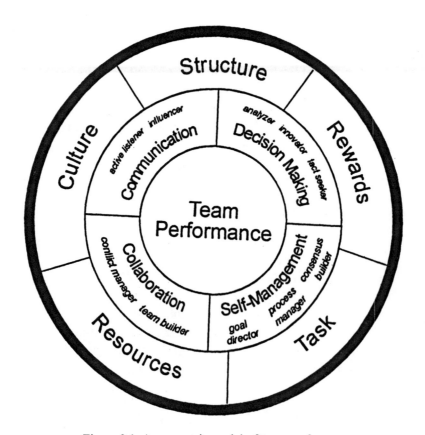

Figure 3.1 A concentric model of team performance

These five variables constitute the organizational environment in which a team works. While the variables are considered outside the sphere of team influence, they can and do have a powerful impact on a team's performance. In effect, they set boundaries on the team. However, it should be noted that these factors in turn can be altered and reshaped by the outcome of a team's performance. Rewards and resources for the next team project, for example, can be increased or decreased. Tasks can be reassigned, and both the culture and structure of an organization can be realigned as a result of outstanding (or poor) team achievement.

While the model illustrates that team performance is strongly influenced by the broader environment in which it operates, it also stresses that team performance depends on the behavior of individual team members. To understand this behavior, we have to examine the next circle of factors in the model (see Figure 3.1). These variables are the ones over which individual team members have more direct influence. By mastering the following four behaviors, team members can transform their resources into meaningful results and truly establish themselves as a cohesive unit:

- **Communication;**
- **Decision Making;**
- **Collaboration; and**
- **Self-management.**

These four key behaviors are as important as any technical expertise team members may have. If they are not frequent and discernable parts of a team's interaction, long-term success is unlikely.

Four Key Behaviors — Ten Key Roles

In this section, each of the above behaviors is defined and the corresponding roles associated with the behavior are described. For teams to become effective, the four behaviors (and 10 roles) need to be performed. Every team member can play each of the 10 roles or different members can play different roles. What is important is that all 10 roles be performed at a satisfactory level (see Figure 3.1).

1. Communication. Good communication involves helping to create and sustain a team environment in which all team members feel free to speak and listen attentively (Campion, Medsker, & Higgs, 1993). Communication experts tell us that 55% of our typical communication in the workplace, school, or office involves listening-related skills. Listening carefully and actively can improve understanding and thus reduce conflict. Listening more and speaking less is good advice in any team effort. As a good communicator, do you listen well without interrupting? Can you restate the problem so as to demonstrate your understanding? Can you express your ideas clearly and concisely? Can you solicit feedback and accept criticism? There are two fundamental roles in communication:

Active Listener — A team member who is very attentive to what others are saying and takes an active role to ensure that what is said is fully understood. An "active listener" provides constructive feedback to other members based on a full understanding of what is heard.

Influencer — An individual who conveys his or her viewpoints in a manner that wins support from fellow team members. This person presents issues in a confident manner and uses facts to get points across to the listener(s).

Table 3.1 presents specific team member behaviors related to effective communication.

Table 3.1 Communication Team Behavior	
Active Listener Role	▪ Listens attentively to others without interrupting ▪ Conveys interest in what others are saying ▪ Provides others with constructive feedback ▪ Restates what has been said to show understanding ▪ Clarifies what others have said to ensure understanding
Influencer Role	▪ Articulates ideas clearly and concisely ▪ Uses facts to get points across to others ▪ Persuades others to adopt a particular point of view ▪ Gives compelling reasons for ideas ▪ Wins support from others

2. Decision Making. Decision making is done best *by* the team, not *for* the team. It involves a clear understanding of the problem or task, gathering and weighing alternatives, achieving consensus whenever possible, and communicating that decision in a timely, acceptable manner (Thompson, 2000). As a team decision-maker, do you seek the input of everyone affected by that decision? Do you carefully consider options and alternatives? Do you have someone play devil's advocate so you see the big picture? Do you encourage a measure of risk taking for innovative solutions? There are three fundamental roles in decision making.

Analyzer — An individual who thinks logically and reviews each situation from several viewpoints. An "analyzer" encourages fellow team members to explore and discuss all alternative solutions before making a final decision.

Innovator — A team member who always generates new ideas and encourages others to do the same. He or she suggests new ways of looking at problems and challenges the way things normally are done.

Fact Seeker — A person who encourages fellow team members to use facts as the basis for all decisions. He or she seeks information from all sources including those "outside" the team.

Table 3.2 presents specific team member behaviors related to decision making.

Table 3.2
Decision Making Team Behaviors

Analyzer Role	<ul><li>Analyzes problems from different points of view</li><li>Anticipates problems and develops contingency plans</li><li>Recognizes the interrelationships among problems and issues</li><li>Reviews solutions from opposing perspectives</li><li>Applies logic in solving problems</li><li>Plays devil's advocate role when needed</li></ul>
Innovator Role	<ul><li>Challenges the way things are being done</li><li>Solicits new ideas from others</li><li>Generates new ideas</li><li>Accepts change</li><li>Suggests new approaches to solving problems</li></ul>
Fact Seeker Role	<ul><li>Offers solutions based on facts rather than "gut feel" or intuition</li><li>Discourages others from rushing to conclusions without facts</li><li>Organizes information into meaningful categories</li><li>Helps others to draw conclusions from the facts</li><li>Brings in information from "outside" sources to help make decisions</li></ul>

3. Collaboration. In many ways, collaboration is the essence of teamwork. It involves working with others in a positive, cooperative, and constructive manner. Collaboration requires that you demonstrate a commitment to the team's overall purpose and to supporting other team members. It requires that team members share responsibility for group functioning and productivity (Wheelan, 1990). As a collaborative team member, are you assisting others on the team? Do you recognize and accept the strengths and weaknesses of other team members? Do you approach problems and conflict openly, without blame? Do you share credit for team successes? There are two fundamental roles in collaboration:

Conflict Manager — A team member who consistently acknowledges issues that the team must confront and resolve. An effective "conflict manager" works to resolve differences of opinion and negotiates solutions that all team members can accept.

Team Builder — An individual who encourages all members to participate in team activities. He or she values and reinforces the contributions of all members. The "team builder" cooperates with others and willingly shares information.

Table 3.3 presents specific team member behaviors related to collaboration.

Table 3.3
Collaboration Team Behaviors

Conflict Manager Role	■ Acknowledges issues that the team needs to confront and resolve ■ Encourages ideas and opinions even when they differ from his/her own ■ Works toward solutions and compromises that are acceptable to all ■ Helps reconcile differences of opinion ■ Accepts criticism openly and non-defensively
Team Builder Role	■ Shares credit for success with others ■ Cooperates with others ■ Encourages participation among all participants ■ Shares information with others ■ Reinforces the contributions of others

---

4. Self-Management. The focus on communication, decision making, and collaboration increasingly move one towards the behavioral category of "self-management" in the team process. It is important to recognize that *management* here is not the exclusive domain of a single individual or layer of bureaucracy; rather, it is a behavior expected of every team member (Wellins, Byham, & Wilson, 1991). A self-manager within a team will collaborate with others in seeking a team solution or new direction. The effective self-manager communicates clearly and effectively, keeping the team on task. As a model of self-management, can you allow yourself to empower others and not feel the necessity to control the work team? Are your professional standards and personal integrity high? Do you have an unwavering commitment to producing a quality product? Are you a self-starter? There are three fundamental roles in self-management.

>*Goal Director* — An individual who helps the team identify its goals and ensures that goals are understood by all members. This person encourages the use of action plans and timetables to help meet team goals.

>*Process Manager* — A person who ensures that the team stays on focus and uses meeting time in an efficient manner. Suggests ways for the team to proceed when needed.

>*Consensus Builder* — A team member who solicits input from all members and encourages them to express their views candidly. The "consensus builder" frequently polls people regarding their current position on an issue and summarizes the team's position.

Table 3.4 presents specific team member behaviors related to self-management.

Table 3.4
Self-Management Team Behaviors

Goal Director Role	▪ Monitors progress to ensure that goals are met ▪ Creates action plans and timetables for work session goals ▪ Defines task priorities for work sessions ▪ Ensures that goals are understood by all ▪ Puts top priority on getting results
Process Manager Role	▪ Stays focused on the task during meetings ▪ Uses meeting time efficiently ▪ Suggests ways to proceed during work sessions ▪ Clarifies roles and responsibilities of others ▪ Reviews progress throughout work sessions
Consensus Builder Role	▪ Solicits input from all members ▪ Encourages frequent polling among team members ▪ Summarizes the team's position on issues ▪ Involves others in decisions that affect them

Each of the above four behaviors and 10 roles is critical to a team's success. As team members acquire the requisite skills needed to perform these roles, a team's performance will improve. However, it takes time for individuals to learn and develop these skills. While some individuals may possess more abilities than others in some areas, it is imperative that all teams have all 10 roles played. Team members can (and should) receive feedback on how well they are performing. In the next chapter, *The Team Developer* is introduced. It provides a formal, non-threatening system for individual members to obtain feedback from other team members regarding their performance on the four key behaviors identified, as well as to what extent they are performing the 10 essential team roles. This feedback will enable each member to develop individual team skills, as well as help the team to perform more effectively overall.

CHAPTER 4

THE TEAM DEVELOPER:
AN ASSESSMENT OF YOUR TEAM BEHAVIOR

A centipede may be perfectly happy without insight or awareness,
But, after all, it restricts itself to crawling under rocks.

Unknown

Obtaining feedback on one's performance is an essential component of growth and development. Try to imagine bowling without seeing how many pins you knocked over! Or playing baseball without seeing where the ball goes after you hit it! Or playing the guitar without hearing the music! Being a member of a student team follows the same line of logic. How do you know where to improve if you do not know how *you presently are performing*? How do you enhance *your team's performance* without measuring various factors that make up team performance?

In this chapter, we will examine the importance of feedback for personal growth and development. Subsequently, we will present an instrument called *The Team Developer* that can be used to obtain feedback from team members. We will discuss how to use this assessment and feedback system and present a sample report. Finally, we will review the mechanics of constructing a development plan you can use to improve as a team member. This plan will enable your individual performance to improve, as well as the overall effectiveness of your team.

The Johari Window

Joseph Luft and Harry Ingham (1963) invented a unique way of examining the amount of personal insight one has about one's own behavior. They referred to the concept as the "Johari Window," after their first names (see Figure 4.1). These authors contend that if you

	Known to self	Unknown to self
Known to others	Open	Blind *(to self)*
Unknown to others	Hidden *(from others)*	Closed

Figure 4.1 The Johari Window

Open	Blind
Hidden	Closed

Figure 4.2 An example of a closed, unhealthy relationship

envision that all of your behavior could be placed in a box, some of it will be known by *you* and some of it will be unknown to you. Furthermore, some of your behavior will be known to *others* and some will not. Luft and Ingham conceptualize this phenomenon in terms of a four-cell matrix.

As can be observed in Figures 4.2 and 4.3, each cell in the Johari Window is affected by how much feedback others give us regarding our behavior. The less feedback others provide, the correspondingly larger our "blindspot." In contrast, the more feedback others give us, the smaller our "blindspot." Likewise, the more candid we are with others concerning our behavior, the more open the relationship. Luft and Ingham assert that, in the beginning, all relationships are rather closed (see Figure 4.2). When teams form, most individuals are rather guarded as to what they communicate to other members (e.g., one member might think another is acting in a domineering fashion but likely won't say it). Likewise, individuals are somewhat reluctant to share their inner feelings about certain issues (e.g., I may resent that the meeting is beginning ten minutes late but won't say so). As we begin to know and understand each other, relationships should become more open. In other words, "healthy" relationships consist of individuals sharing feedback — being willing to provide and receive behavioral feedback freely and often (see Figure 4.3).

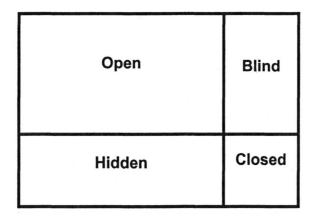

Figure 4.3 An example of an open, healthy relationship

If I am unaware that you are offended by my behavior, I am not likely to change. If you tell me what your needs are and whether I am meeting them, I am more likely to satisfy them. Consequently, "healthy" teams are much more likely to grow and develop, because members converse openly and frequently. "Unhealthy" teams are much more likely to endure unmet needs in private.

Table 4.1 identifies some principles of personal insight. Overall, they stress the importance of viewing feedback as a gift. A gift that we as teammates must learn to accept and give. A gift that we must receive in order to grow and develop as individual team members. A gift that we must receive to grow and develop as a student team.

Table 4.1
Principles of Personal Insight

1. A change in any one area of the Johari Window will affect all other areas. By increasing or decreasing the amount of *feedback* we give and ask for, we alter our relationship with others.
2. It takes energy to hide, deny, or stay blind to behavior involved in team interaction.
3. Threats tend to decrease insight; mutual trust tends to increase insight (i.e., if you trust the other person, feedback levels are increased).
4. The smaller the upper-left area, the poorer the communication and less effective the team's performance.
5. Learning about team processes can help increase insight for the team as a whole, as well as for individual members.

The Team Developer is an assessment and feedback system that makes the process of receiving and giving feedback in teams relatively painless. It focuses on the four behaviors and 10 roles outlined in the previous chapter, listed below in Table 4.2. In the next section, we will describe this instrument and review how to use it in your team.

Table 4.2
Critical Behaviors and Roles Individuals Can Perform in Teams

Behaviors	*Roles*
Communication	➢ Active Listener
	➢ Influencer
Decision Making	➢ Analyzer
	➢ Innovator
	➢ Fact Seeker
Collaboration	➢ Conflict Manager
	➢ Team Builder
Self-Management	➢ Goal Director
	➢ Process Manager
	➢ Consensus Builder

The Team Developer

The Team Developer is a computer-based survey consisting of numerous questions that relate to the four behavioral categories: (a) communication, (b) decision making, (c) collaboration, and (d) self-management. Team members use a 1- to 5-point rating scale to anonymously rate themselves and all of their fellow team members on various team behaviors. Based upon those ratings, each team member receives a confidential feedback report. The report summarizes how each person sees himself or herself, as well as how their team members perceive them.

The instrument is administered by providing *each* team member with a computer disk containing the survey items and the names of all team members. The disks are prepared by the course instructor (or a teaching assistant) who enters the individuals' names and team affiliation into a data file and then copies this file onto disks. When team members access the file on the disk, they select their name from a list that in turn indicates the team of which they are a member. The program prompts the student to rate themselves and each of their team members on every survey item. When finished, team members return the disks to their course instructor who uses the *Team Developer Administrator's Program* to analyze survey results and generate feedback reports.

The Team Developer has been used by thousands of undergraduate and graduate students during the past several years. Student teams enrolled in engineering design courses, business simulation classes, organizational behavioral classes, and group dynamics courses have used the instrument. Universities such as Ohio State, Stevens Institute of Technology, and the New Jersey Institute of Technology have used *The Team Developer*. In many cases, the survey has been administered twice during a semester — once at the midpoint and again at the end of the term. This approach permits a longitudinal assessment to ascertain whether teams (and its individual members) are progressing as the semester unfolds.

Constructing a Personal Development Plan

Confucius wrote over two thousand years ago, "Only the supremely wise and the abysmally ignorant do not change." We suspect you are neither. A critical component of *The Team Developer* is the construction of an individualized action plan for improvement. Based on the feedback you receive from your teammates, you should carefully prepare a series of steps (i.e., a development plan) to enhance your team skills. This plan will enable you to grow the interpersonal and team-related skills that are so much in demand in today's workplace. Unless you make a concerted effort to understand the results and focus on improvement, *The Team Developer* system will not benefit you.

The *Team Developer Feedback Report* includes a planning sheet that you can use to address the developmental areas identified by you and your peers (see Table 4.3 for an illustration). To use the planning sheet, describe the developmental area(s) you want to focus on in the first column. Rather than working on an entire dimension all at once, try to select developmental areas related to specific survey items. For example, instead of focusing on decision making, closely investigate those specific items of decision making where you are rated lowest. In the second column, list the action steps you plan to take to improve your performance. You particularly want to examine the feedback for areas of agreement. Where do you *and* your teammates *both* view you as weak? What specific items are rated lowest? Although sharing your results with others in your team is not required, obtaining additional input from other team members can help you prepare more effective action steps. Perhaps, you can choose a team member that you are close with for a one-on-one sharing session. The final step in your development plan is to record your target completion dates in the third column.

Table 4.3
Personal Development Plan

Development Area	Action Steps	Completion Date

You will have more success in improving your performance if your action steps define *precisely* what you are going to do. You want to avoid vague, general actions that are difficult to measure. With clear action steps, you can determine whether you have met, exceeded, or fallen below your objectives. It also helps to have more than one action step for a particular developmental area. Examples of vague and specific action steps are shown in Table 4.4 to guide you as you write your own personal development plan.

Team members (and consequently the whole team) will benefit greatly by sharing individual feedback reports and action plans with each other. The feedback provided by *The Team Developer* will enable your team to discuss overall patterns of member results to determine whether various team members are meeting all 10 team roles. If not, a team might deliberately decide to assign specific roles to certain team members. Obviously, this discussion should be incorporated into individual member action planning. From a team perspective, it is necessary that all 10 roles be performed. However, it is not necessary that all members of the team individually perform all 10 roles.

Table 4.4
Example of a Completed Personal Development Plan

Development Area	Action Steps	Completion Date
Persuading others to adopt my views.	***Vague:*** *Present better. Think out arguments in advance of each meeting.*	*10/15*
	Specific: *(a) Prepare in advance a fact-based argument to support my position on XYZ issue. Seek time to present my position at our next meeting.*	*10/15*
	(b) After the meeting, seek feedback from two team members on the impact of my argument.	*10/15*
	(c) Read the book, How to Sell Your Ideas.	*10/30*
Generating alternative solutions.	***Vague:*** *Be more creative by thinking about things from different perspectives and talking to people from other backgrounds.*	*11/30*
	Specific: *(a) Prior to making a decision on XYZ, seek the input of at least three people whose backgrounds are different from one another and from my own.*	*11/15*
	(b) Keep a notebook handy to write down new ideas. Review my ideas once a week with a view toward how they will help our team solve problems.	*11/30*

CHAPTER 5

THE ART OF GIVING AND RECEIVING FEEDBACK

If you do what you've always done,
*   you're gonna get what you always got!*

Yogi Berra
Baseball Player and American Philosopher

Art Garfunkel sang a very popular song back in the 1970s with the lyrics, "I bruise you, you bruise me. We both bruise too easily." Giving and receiving feedback can be a very tricky proposition. It seems like people can get upset (bruised) so quickly if they are told something that they do not want to hear. Yet, individuals need to be informed of performance problems or they will continue to behave poorly. Unless we realize there is a problem, we cannot correct it. When a team member performs poorly, it is critical that we let him or her know. Or, in the words of Yogi Berra, "you're gonna get what you always got."

The question becomes "how do you communicate the problem to them?" On the other hand, if *you* are the team member who is performing poorly, "how can others tell you without making *you* defensive?" In this chapter, we will review some guidelines for *giving* and *receiving* feedback effectively. We also will present some tips on good listening.

The "How To" of Giving Feedback

Feedback is the process of communicating to individuals information about their behavior or performance on a task; it can occur one-on-one or in a group setting (Schein, 1988). The purpose of feedback should be to improve performance by openly addressing individual and/or team problems. One indication of an effective team is the ability of its members to give and receive feedback constructively. Nonetheless, the process is seldom easy. Team members risk exposing themselves to a variety of reactions when providing feedback to other members — resentment, hurt feelings, embarrassment, anger! However, with some forethought and planning, feedback can help to build trust among teammates and can lead to rewarding experiences for everyone involved.

All of us must realize that providing effective feedback to fellow team members requires tact, diplomacy, and an environment of mutual openness and respect. All members need to view feedback as a way of giving help. Unless individuals are made aware how you perceive their actions, they are likely to continue them. For example, John is consistently a few minutes late for your team meetings. It is becoming a bigger and bigger irritation to you. Unless you inform John how you feel, he cannot satisfy your needs. The key piece in the feedback puzzle is to convey the information in such a manner that it is viewed by the member as a "way of giving help."

Table 5.1 presents several characteristics of effective feedback. Each of these factors can help you structure your feedback to team members. Whether you are completing *The Team Developer* itself, interacting one-on-one with another individual during a feedback discussion session, or simply sharing your perceptions in a team meeting, it is vital that you give feedback that will be constructive and developmental in nature.

Table 5.1
Characteristics of Effective Feedback

Giving feedback is most effective when:
- Both parties (receiver and giver) view it as a way of providing help.
- It is direct and honest; clearly communicating the problem that needs addressing.
- You focus on specific issues and behaviors.
- You provide specific examples and performance incidents.
- You cite positive information *first*, then the negative.
- You identify only two or three key areas for improvement.
- The recipient is involved in developing solutions.
- It is reciprocated.
- It gradually moves to a deeper level (e.g., like peeling an onion).
- It is well-timed.
- It is checked to ensure clear communication has occurred.
- Performance expectations are known in advance (i.e., "feed-forward" your expectations).

Feedback should be direct and honest. If feedback is going to lead to improvement, the receiver has to understand the problem. That is, the giver has to define his or her perception of the problem as clearly and honestly as possible. Returning to the previous example of John's tardiness at meetings, let us assume another team member (Sara) is getting anxious, because she believes that the team's work cannot be completed if he continues to arrive late. If she tells John, "I don't want you to arrive late anymore," she is really being indirect, telling him only part of the problem. Sara is much more likely to have an impact if she explains to John, "I am concerned that your arriving late will prevent us from getting all of our work done on time." The advantage of the second approach is that John now has an understanding of how his behavior may effect the team's performance. Being direct also involves acknowledging and communicating how another's behavior affects members' feelings. When Sara told John that she was "concerned," she was communicating her feelings about the situation. This information let John realize that the problem was not just his lateness, but how it was making her feel. Framing the problem this way is less likely to engender a defensive reaction.

Feedback should be specific and behavioral in nature. Another important aspect of effective feedback is making sure that it is specific and stated in behavioral terms. This approach means basing it on observation and not on inference or judgement. In addition, the feedback should be related to a specific situation and behavior rather than a general, conceptual, or poorly defined trait. For example, feedback may be stated conceptually as in (a) s/he lacks business sense, (b) s/he has a bad attitude, (c) s/he lacks ambition, or (d) s/he is a poor communicator. These generalizations or global descriptions may seem like an economic form of communication, but they likely will lead to confusion or misunderstanding by the receiver. Further, negative generalizations may lower a person's sense of worth and self-esteem. The impression given is that the individual (not the behavior) is the problem. Moreover, the receiver typically does not know what to do differently when presented with generalized feedback that lacks behavioral definition.

Feedback should be timely. One of the best ways to keep feedback specific is to give it immediately after a behavior has occurred. Keeping feedback recent probably is the best way to ensure that people will understand what you are talking about. It also enables a team to resolve problems before they worsen. In reality, giving immediate feedback may not always be practical. People may not realize something is a problem until sometime later. On other occasions, a problem may be more complex than the team currently has time to devote to it. Moreover, the team may not yet possess the skills to provide effective feedback. In such situations, videotape is an excellent way to retain a visual record for later discussion. At other times, if an issue is particularly sensitive or personal, a team member may decide it is best to speak with an individual in private. When providing feedback to an individual on a one-on-one basis, make sure that you are relatively free from interruptions and distractions.

Use the "feed-forward" technique of giving feedback. We all have heard of the word, "feedback." It pertains to a situation where an individual does something, and subsequently we *feed back* our views of that behavior. The sequence of events is behavior, then evaluation. With "feed-forward," we share our expectations first and make sure the individual knows what we desire. Subsequently, we evaluate them on that expectation. Frequently, people perform behaviors without realizing what we expect. Teams generally are more successful when members recognize what the ground rules are up-front. In other words, members are informed before the fact what other members or the course instructor expect.

The "How To" of Receiving Feedback

When someone gives you feedback via *The Team Developer* or in person, you have a choice. You can deny it — that is not true; that's not how it happened! You can make excuses — I would have been here on time if the bus wasn't late! You can get angry. You can get embarrassed. All those reactions are not very healthy and likely will result in no behavioral improvement. On the other hand, we can approach this feedback from a healthy perspective. We can recognize that no one is perfect, including ourselves. We can view the feedback as an opportunity to grow and improve. We can attempt to understand why the individual feels the way s/he does and reexamine the data from his or her perspective. Finally, we can deliberately develop action plans for improving our performance. Hence, we can approach the feedback in terms of *learning >>> growing >>> developing >>> evolving.*

In Table 5.2, we present some of the key characteristics of how to receive feedback. Your team members can be a valuable resource in your professional development. Use them wisely. Take advantage of the feedback they are offering you. Listen with a receptive ear. Make plans to change behaviors and develop your team skills.

Table 5.2
Characteristics of Effective Feedback

Receiving feedback is most effective when:
- You accept the feedback as "reality" for the person giving it.
- You probe for additional information and examples.
- You focus on how the feedback can help solve a specific problem.
- You summarize what you think has been said to assure understanding.
- You express appreciation for others' input into your growth.
- You are committed to improvement.
- You solicit it regularly.

Listening — A Critical Skill in the Feedback Process

Obviously, listening is an important aspect when *giving* or *receiving* feedback. An individual must be attentive to what others are saying, then process this information accurately. Hearing is an integral component of listening. However, you also need to be very sensitive to a person's nonverbal cues in the feedback process. The communication literature consistently has found that nonverbal behaviors are of equal (or greater) importance in person perception than the (verbal) message itself (De Meuse & Erffmeyer, 1994). Consequently, as you "listen" to what other team members are communicating, pay attention to both *what* they are saying and *how* they are saying it.

Keith Davis (1977) formulated a list of guidelines that he entitled, "The Ten Commandments of Good Listening." These commandments are presented for your information in Table 5.3. We believe they will provide you some sound counsel as you develop your skills to give and receive feedback effectively. In the next chapter, we explore various approaches to managing interpersonal conflict.

Table 5.3
The Ten Commandments of Good Listening

1. Stop talking! You cannot listen if you are talking.
2. Put the talker at ease. Help the talker feel that s/he is free to talk.
3. Show the talker that you want to listen. Look and act interested. Do not read your mail or stare out the window while s/he is talking.
4. Remove distractions. Do not doodle, tap, or shuffle papers.
5. Empathize with the talker. Try to put yourself in the talker's place so you can see that point of view.
6. Be patient. Allow plenty of time. Do not interrupt.
7. Hold your temper. An angry team member gets the wrong meaning from words.
8. Go easy on argument and criticism. This puts the talker on the defensive. Do not argue. Even if you win, you lose.
9. Ask questions. This encourages the talker and shows you are listening.
10. Stop talking! This is the first and last commandment because all the others depend on it.

Note. This table was adapted from Keith Davis (1977), p.387.

CHAPTER 6

HOW TO RESOLVE GROUP CONFLICT

It's easy to get good players. Getting 'em to play together; that's the hard part."

Casey Stengel
Major League Baseball Player and Coach

"That is not what I said," Nick asserted!

"Yes, it is! You always feel we should do what you want. Well, this time you're wrong. This time we are doing it my way," Chris retorted.

"Guys, just wait a minute. Why must one of you believe you are always right? How about we call a truce and discuss the issue at hand?"

Another attempt on your part to ease tempers. It seems as though whenever your group gets together you spend half the time arguing. Why can't your group stick to the issues and have a meaningful, *non*-emotional discussion? It is not a matter of being right or wrong. It is a matter of describing the situation at hand, suggesting alternative ways of addressing the problems, and arriving at a well thought-out decision that represents the combined efforts of the entire group. Isn't that what synergy is all about?

Figure 6.1 Interpersonal conflict can cause many team problems when not managed constructively.

Few people look forward to being in conflict with others. Conflict has the potential of causing interpersonal tension and stress. It can make us feel frustrated, angry, embarrassed, isolated, or perhaps even fearful of being harmed. However, most of us recognize that conflict is inevitable. In fact, not only is it inevitable, but it is normal, natural, and even essential for effective group dynamics.

It would be unhealthy and unproductive to avoid interpersonal conflict altogether. As we mentioned earlier, a team's developmental journey includes going through a conflict phase (Stage 2). Teams that attempt to circumvent this stage usually fail to accomplish their goals. After all, one of the key advantages of a team is that it brings together individuals with diverse perspectives, needs, personalities, values, likes, and dislikes.

When properly managed, conflict can become a great resource. The appropriate team environment can encourage members to think creatively, develop new ideas, and work harder to understand each other. *Constructive* conflict can enhance the quality of decision making and actually increase the overall cohesiveness of the team. Unfortunately, when conflict is not dealt with effectively, interpersonal relationships are damaged, communication breaks down, stress goes up, and group performance wanes. Consequently, it is important to ensure that you and your team members know how to deal with conflict constructively. In this chapter, we will examine three different forms of conflict, identify five ways to manage conflict in teams, provide some signs to help you diagnose how your team deals with conflict, and offer some guidelines on how to resolve conflict effectively when it arises in your team.

Types of Conflict

Most team conflicts fall into one of three broad categories. The first type relates to "personality" or "behavioral style differences." Individual differences are inherent in teams and you will find that some members may be easier to work with than others. People may look at the world differently than you or have differences in terms of how they approach their work. The key is to remain respectful of one another's work styles. It is appropriate to take issue with what someone does or how they do it, but you should avoid attacking that person by labeling or stereotyping them. This approach tends to increase tension and seldom lets others know what needs to change.

Another type of conflict can be categorized as "interpretative differences." These conflicts reflect differences about procedures that should be followed, conclusions reached, facts that are important, or theories that should be considered. Such differences are inevitable and, in fact, vital for effective group functioning. When facing this type of conflict, keep the emphasis on the facts. Maintain a problem-solving orientation and avoid bringing personal characteristics into the discussion.

"Interest-based conflicts" have to do with differences in underlying needs, values, goals, and/or access to resources. One of the keys to solving this type of conflict is searching for common ground upon which to build eventual solutions. Although you may have some divergent interests, it is likely that you share some common interests as well. By focusing on them first, you will be more likely to create a cooperative tone for resolving differences. A common mistake when faced with interest-based conflicts is the tendency to confuse interest with position. While an interest is an underlying concern or issue, a position is a particular stance in relation to the concern or issue.

An example can help illustrate the distinction. Susan asked one of her team members, Jose, for assistance when preparing her part of the report that the team needed to submit. Jose quickly responded that he would not help her. Susan left, feeling frustrated and angry over Jose's lack of cooperation. Jose's *position* was that he would not help Susan. However, what were the *interests* or issues upon which the position was based? If the two members would have continued to explore the conflict by trying to identify underlying interests, they might have been able to reach a mutually beneficial solution. Here are some of the things they could have realized.

- Because they were members of the same team, both shared an interest in ensuring the report was the best it could be.
- Jose's underlying interest was in ensuring that he had enough time to complete his own portion of the report. He worried that if he helped Susan he would not have enough time left.
- Susan's underlying interest was not time, but in ensuring that she include the correct information in the report. Jose, she believed, would be a good source to answer her questions.

Once they had identified each other's interests, at least two solutions would have become apparent. One alternative would be for Susan to help Jose complete his work, so that he would not fall behind after giving her some of his time. A second alternative would be that Jose refer Susan to another person who was equally knowledgeable. By pursuing either of these alternative solutions, their positions would have shifted but their respective interests would have been preserved.

Conflict Management Styles

Another way to think about conflict is in terms of the style or approach that individuals take to resolve it. Thomas and Kilman (1974) identified five general styles. They include (a) avoiding, (b) accommodating, (c) competing, (d) compromising, and (e) collaborating. Each style can be defined in relation to two general dimensions — assertiveness and cooperation. *Assertiveness* reflects a member's willingness to act on behalf of his or her views, interpretations, or interests. It is the extent to which the individual attempts to satisfy his or her own needs. On the other hand, *cooperation* represents a member's need to get along with others and/or avoid conflict. It is the extent to which the individual attempts to satisfy the other person's needs. All five of these conflict styles are presented in Figure 6.2. Each of the styles is defined on the following pages.

Avoiding. Teams or individuals that tend toward avoidance are low in cooperation and likewise low in terms of assertiveness. The primary problem with this style of conflict management is that conflicts are never really resolved. Team members simply do not bring them up. Eventually, it may become difficult for a team to make progress, because members never directly address the real issues. Members simply put off interpersonal differences, pretending that they do not exist in the first place or consciously deciding it is not worth the effort to deal with them.

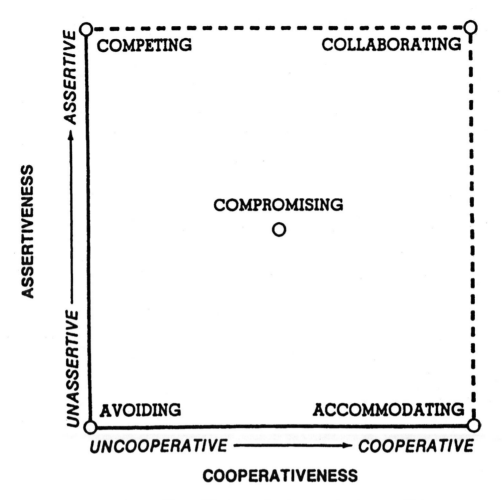

Figure 6.2 Approaches to managing team conflict

Accommodating. This style occurs when cooperation is high but assertiveness is low (see Figure 6.2). "Accommodators" fail to meet their own needs in order to satisfy other members' needs and avoid conflict. Unfortunately, the tendency to be too accommodating often results in poor team decision making. Individuals are so concerned about the feelings of fellow team members that the best solutions are never fully considered. Further, team members who continuously acquiesce may eventually grow resentful. They will likely come to a point where they explode their emotions on the team.

Competing. The competing style reflects being high in assertiveness and low in cooperation (see Figure 6.2). This approach often creates a contentious environment in which conflicts become escalated. Individuals get locked into satisfying only their own needs. Eventually, team members get so agitated that the team's goals are often neglected.

Compromising. On the surface, compromising may seem like a reasonable approach. It implies being moderate in terms of both assertiveness and cooperation (see Figure 6.2). However, compromising also implies splitting the difference or meeting in the middle. In many cases, compromise results in a solution that is the least common denominator and the *least* effective. Frequently, the best answer to a problem involves candid discussion, ongoing critique, and the most direct approach regardless of whose idea it might be.

Collaborating. This style involves being high in terms of both assertiveness and cooperation. Members seek to maximize each other's needs, such as in a "win-win" scenario. Consequently, collaboration is the style most likely to produce optimal and creative resolutions to conflicts. Along the way, team members will build trust in each other and increase their confidence in getting the job performed properly. Collaborators feel comfortable expressing their interests, but want to ensure that other members get their needs met as well.

Diagnosing How Your Team Deals with Conflict

Take a few minutes and attempt to determine which of these five styles best characterizes the way your team handles conflict. After you decide, share your views with other members on the team. Discuss the reasons for your beliefs. Subsequently, discuss the advantages of moving toward a collaborative style to resolving team conflicts. Table 6.1 will provide some information on each style to help in your diagnosis.

Table 6.1
Diagnosing How Your Team Deals with Team Conflict

Some clues that your team may be "avoiding" conflict:
- Do team members usually accept ideas and solutions without thoroughly discussing their pros and cons?
- Do team meetings end without members having a clear idea about what happened or what should occur next?
- Do the same problems or issues keep coming up over and over?
- Do meetings typically start later than planned and/or is there a tendency to get off the subject at once?

Some clues that your team may be too quick to "accommodate:"
- Do members tend to quickly back down on their positions?
- Would you say members on your team are overly polite to one another?
- Do you and/or others feel uncomfortable saying what you really think or feel?
- Do one or two people dominate discussions and action planning?
- Would your team say that keeping everyone happy is more important than finding the best solution to a problem?

Some clues that your team may be in a "competing" mode:
- Do members frequently pass blame on others when things do not go as hoped or planned?
- Can you identify specific cliques or subgroups within your team that always seem to stick together on every issue?
- Is there a tendency to label or stereotype individual members?
- Are members usually reluctant to consider alternatives other than their own positions or ideas?

Table 6.1 (continued)
Diagnosing How Your Team Deals with Team Conflict

Some clues that your team may be in a "competing" mode: (continued)

- Do members get you agitated easily (e.g., if a discussion lasts longer than expected)?
- Do members frequently interrupt or talk while others are speaking?
- Do members lecture one another to convince them that they are right?

Some clues that your team may be too quick to "compromise:"

- Do you settle most differences of opinion by voting?
- When looking for solutions to problems, would you say your team focuses mostly on incorporating everyone's position (as opposed to trying to identify the best possible solution)?
- Is there a general lack of enthusiasm amongst team members for the ways issues are resolved or problems are solved?
- Does your team avoid vigorous debate on issues or topics?
- Would you conclude that most of the solutions your team reaches are less than ideal?
- Do you sense an overall lack of commitment to team decisions?

Guidelines for Transitioning to a Collaborative Style

The following guidelines will help your team transition to a collaborative approach to resolving conflict:

1. **Frame conflict in terms of problems to be resolved, rather than by focusing on who should be blamed or held responsible.**
2. **Develop objective criteria that will help your team evaluate the merits of proposed solutions to conflicts.**
3. **Consider several alternative solutions and look for ways to combine alternatives in order to develop an ideal solution.**
4. **Actively listen to one another, taking time to convey an understanding of each other's points of view.**
5. **Develop clear follow-up plans to ensure resolutions.**
6. **Stress the goals and perspectives in *common* among team members.**
7. **Seek compromise solutions only as a very *last* resort.**
8. **Encourage, support, and reward candid and frequent communication among team members.**

CHAPTER 7

HOW TO CONDUCT EFFECTIVE TEAM MEETINGS

Courage is what it takes to stand up and speak.
Courage is also what it takes to sit down and listen.

Winston Churchill
Former Prime Minister of England and Statesman

Do you spend too much time with your team in meetings? Are the meetings you attend too long? Do you often question what really was accomplished during the meetings? Do you frequently get sidetracked? Do some individuals have something to say about everything, while others rarely say anything? Do your meetings tend to start 10 or 15 minutes after the agreed upon start time? If you are like most students, you have answered "yes" to most of these questions.

Successful meetings do not just happen; conducting them is not an easy task! Some meetings are not even necessary. Perhaps, nothing is more discouraging than to hold a meeting when memo, e-mail, or a telephone call could have easily handled the business. Equally discouraging is a meeting that starts late or runs too long. Team meetings require hard work and much skill on the part of all members. As in the words of Winston Churchill, members must have the courage (and knowledge) to assert an important point when necessary versus being quiet and listening. In this chapter, we will examine the symptoms of poor meetings, various roles members should play during meetings and guidelines for facilitating effective meetings. Also, we will review some useful tips on how to handle conflict during a meeting. Specific examples of a meeting of design engineering students facilitated ineffectively and effectively are presented.

Symptoms of Poor Team Meetings

There are several signs that can tell you that your meetings are not going well. However, you need to know what to look for and what needs to be corrected. Perhaps, it is the meeting structure that's the problem. Perhaps, it is that team members are not assuming their roles. Perhaps, it is that meetings have no agenda, end time, or problem-solving process to examine important questions. Deborah Harrington-Mackin (1994) identified many symptoms or warning signs of internal meeting problems. Some of these key symptoms are presented in Table 7.1.

Overall, two of the primary indicators that suggest team meetings are being run ineffectively pertain to member respect and responsibility. Today's student is extremely busy. Most students have to balance the demands of a job, classes, and a relationship. Finding time to attend meetings with fellow students outside of class is very difficult.

When team meetings are set, do all students show up? Do students arrive on time? Do they leave early? Furthermore, do *all* students contribute during meetings? Are they prepared to discuss the topic at hand? Do they respectfully listen to each other's viewpoints? Do they volunteer for follow-up activities? Do the members of the team feel "like a team" following the meeting?

Table 7.1
Symptoms of Internal Meeting Problems

1. Team meetings generally begin 5 – 15 minutes late.
2. Members often arrive late, leave early, or never even show up for the meetings.
3. No agenda exists — students simply have a vague notion what they want to accomplish.
4. One or two members monopolize discussion throughout the meeting.
5. Members have not read the assignment, performed the necessary background research, or did what they were expected to do. Consequently, individuals are ill prepared for the meeting.
6. Non-verbally (or verbally) some members clearly convey that they would rather be elsewhere.
7. Members constantly interrupt each other or talk in pairs without listening to the individual who has the floor.
8. Additional discussion occurs in "pairs" or "triangles" after the meeting supposedly has ended.
9. Issues never get resolved, only put on the back-burner until next time.
10. No follow-up action plan is developed. Members are confused with regard to what is the next step and who is responsible for performing it.
11. The same individual or individuals end up doing the majority of the work.
12. The meetings run on and on and on.
13. Assignments are not completed on time or are completed poorly.

Note. This table was adapted from Harrington-Mackin (1994).

Member Roles during Team Meetings

Meetings will run more smoothly when the following four roles are assumed by various members of the team: (a) primary facilitator, (b) scribe, (c) timekeeper, and (d) secondary facilitator. The "primary meeting facilitator" is responsible for activities *both* prior to and during the meeting. *Prior to the meeting*, this member should state the purpose of the meeting, develop the agenda, state the desired outcomes from the meeting, and assign the roles of scribe and timekeeper. *During the meeting*, the primary facilitator is responsible for beginning the discussion, soliciting input from all team members, processing this information, helping the team draw conclusions and develop action plans, as well as ensuring that the meeting stays on track. In addition, this individual likewise should be an active contributor of ideas. However, the primary facilitator must be careful when s/he desires to contribute. It is a good approach to inform the other team members that you are changing roles. For example, "During the next minute I will put down my primary facilitator hat and share some thoughts that I have on this topic" (Kayser, 1995).

The roles of scribe and timekeeper also are very important to the success of a meeting. The "scribe" has the responsibility of taking notes during the meeting. This member provides feedback to the team on what is being recorded and checks for consensus. After the meeting, the scribe should send members a follow-up memo of the notes. The "time keeper's" job is to ensure that the meeting remains on schedule. When members begin to focus on tangential issues, the timekeeper can assist the primary facilitator to assure the meeting stays on track. We highly recommend that team members switch roles from meeting to meeting in order for all members to develop these skills.

The fourth role is entitled "secondary facilitator." It is everyone's responsibility to assume this role. Secondary facilitators come to the meeting prepared, actively contribute during the meeting, and freely volunteer for assignments following the meeting. When members actively play these four roles, meetings (and teams) are much more successful.

Guidelines for Facilitating Effective Team Meetings

The role of "primary facilitator" is absolutely critical to the effectiveness of a meeting. A skilled facilitator can ask just the right question, can draw out an idea from a member at just the right time, and can make sure an "under-developed" concept is not prematurely dismissed. However, students typically lack the leadership experience to facilitate meetings. They have the ability but lack the knowledge and practice to be effective facilitators. The more you practice, the more comfortable you will become facilitating. We have provided some guidelines below to help you facilitate team meetings.

Remember that, in order to improve, you must take the initiative to practice these skills.

1. **Start the meeting on time. Do not keep three responsible members waiting, because one *irresponsible* one is late. If a member is late, it is his or her responsibility to find out what was missed.**
2. **Ask members for their opinions and feelings to encourage discussion.**
3. **Paraphrase what is said to help members understand each other.**
4. **Ask for specific examples to enhance comprehension.**
5. **Clarify member assumptions. For example, "Your point assumes that . . . . . Is that right?"**
6. **Probe an idea in greater depth to solicit more insight.**
7. **Summarize key points to ensure understanding.**
8. **Encourage open-mindedness among members.**
9. **Reflect on what someone is feeling or expressing.**
10. **Refocus the team back to the issue at hand (when needed).**

Guidelines (continued)

11. **Surface differences of opinion among team members.**
12. **Promote a new way of thinking about an issue. For example, "Let's pretend you are the manager for a minute. How would you feel?"**
13. **Recommend a process on how to move forward. For example, "I suggest that we brainstorm as to why the customer said . . . ."**
14. **Test team members for consensus.**
15. **Move the team toward making a decision and/or action planning.**

How to Handle Conflict During Team Meetings

One of the biggest challenges for team members is how to handle conflict during meetings. Whenever you have two or more individuals together, you have the potential for conflict. Team members may have differences of opinion on almost anything (e.g., what the relevant facts of the case are, how to satisfy the customer's needs, should we address point one or two first). As we previously asserted, it is important for the team to recognize that interpersonal conflict is a normal part of team dynamics. Nevertheless, how a team processes conflict is extremely important.

When conflict occurs during a team meeting, the "primary facilitator" should deal with it immediately. The first action for the facilitator is to take a time out. During this minute of silence, tempers need to calm down. Request that everyone jot down their thoughts on the problem area. After the period of silence, invite the members to share their *feelings* on the problem. Subsequently, have the members share what they consider the *facts* regarding the problem. It is important that you process feelings first, then facts. Individuals are emotional. To pretend that those emotions / feelings do not exist or are irrelevant is unwise. Indeed, member feelings are equally as important as facts. Both need to be identified, recognized, and processed. Once team members discuss their feelings and the facts of the case, solutions can be proposed. Overall, when facilitating team conflict resolution, the sequence is one of:

FEELINGS

⇓

FACTS

⇓

SOLUTIONS

CHAPTER 7: HOW TO CONDUCT EFFECTIVE TEAM MEETINGS

SCENE ONE: THE *WRONG* WAY TO CONDUCT A MEETING

A student group has gotten together to discuss an assignment for a design engineering class. The group previously had developed a conceptual drawing for a new fuel efficient gas engine. Greg and Mike were given the responsibility of investigating what materials could be used to construct a prototype. Jessica and Kamomo agreed to examine various engines currently in the marketplace and ascertain how their model could compete. Babu and Sally were charged with exploring government regulations and laws that might be germane in the manufacturing process. All six members of the team arrive at the designated meeting time. No one has been assigned roles to facilitate the meeting.

Mike: Well, it looks like everyone is here. Babu, it was nice that you could make it this time!

Babu: I'll pretend I didn't hear that. Hey! Do you know whether the Brewers won last night? I know in the fifth they were leading the Braves by two runs.

Kamomo: Sally, were you at class yesterday? Could I borrow your notes?

Sally: Of course you can. Although to be honest with you, you really didn't miss much.

Mike: Ok. What about if we get down to business. Kamomo and Jessica, you were supposed to look at possible competitors of our engine. What did you find?

Jessica: Well...Kamomo and I didn't get around to doing it yet. We both had a major test in Biology. We needed to spend all of our free time preparing for it.

Kamomo: Yeah. We should be able to get it by this weekend.

Greg: What do you mean, this weekend! You were supposed to have it done by today. How can we accomplish our class project if you don't do your part?

Jessica: Let's discuss what materials we can use to construct the prototype during this meeting. Mike and Greg, what did you find?

Mike: Well, we didn't have much time. We only...

Sally: What?! You're not finished either! Why? You said that...

Babu: I got to go. I have a History exam at 3:00 and I have got to review my notes one more time. Let me know when our next meeting is, ok? See you around. (Babu leaves the meeting.)

Greg: Here we are again. What should we do now?

SCENE TWO: THE *RIGHT* WAY TO CONDUCT A MEETING

The same set-up as in Scene One. This time Mike has been given the role of primary facilitator. Sally is timekeeper and Kamomo is scribe.

Mike: It's 2:00; let's begin. First, thanks to everyone for being here on time. As our agenda states, our objectives today are (a) examine what materials can be used to construct our prototype, (b) explore other engines in the marketplace which might compete with our model, and (c) investigate any pertinent governmental laws that we should be aware of. Is that right? Any questions or comments? If all goes as planned, the meeting should be over by 3:30.

Kamomo: Sally, were you at class yesterday? Could I borrow your notes?

Mike: Kamomo, how about you and Sally discuss that issue after our meeting? Let's focus on why the team is here today. Babu and Sally, in your research of laws and governmental regulations, what did you find?

Sally: Based on our findings, it appears the EPA needs to be considered. The Environmental Protection Agency is extremely concerned about any pollution which might be caused in the manufacturing of our engine. The regulations are very clear and...

Jessica: Come on! Who cares about those "tree-huggers" any way? If the government had its way, we would still be back in the horse and buggy days.

Greg: Jessica is right. We can always get around those guidelines anyway. Let's focus on how we can make the most cost-effective engine.

Babu: Wait a minute. It is up to all of us to take care of the world. If tomorrow's business leaders don't, there won't be a need for our engine. There won't be an Earth.

Mike: Wait a minute. Let's calm down. These are important issues. However, we need to discuss them rationally and carefully. Let's take a moment of silence. During this time, everyone jot down their thoughts about the environment, our prototype, and how we potentially could manufacture it. After we share our *feelings*, we can proceed by examining the *facts* of this case. Finally, we will develop a list of *solutions* based on our collective input. Now, let's put our heads down and gather our thoughts. Sally, let us know when three minutes are up.

CHAPTER 8

BECOMING AN EFFECTIVE TEAM MEMBER

"Come to the edge of the cliff," he said.
"We're afraid," they said.
"Come to the edge of the cliff," he said.
"We're afraid," they said.
"Come to the edge of the cliff," he said.
      They came.
        He pushed.
          They flew!

        Guillaume Appollinaire
        Author

Becoming an effective team member takes work! It takes time, effort, and practice to attain team skills. And you will make mistakes. You will encounter situations that make you uncomfortable. You may fail at times. However, the acquisition of any skill set is a journey. Remember when you first learned to drive a car. Every moment behind the wheel required concentration and diligence. Then there was parallel parking. That required a skill it seemed like you would never attain. But you did!

The same is true with successful team performance. Team performance requires a new set of skills. You may be a superb individual contributor, receiving an "A" on all individual assignments. On the other hand, performing well in groups is different. You may earn only a "C" on group assignments. Why even try? It would be much easier to bury your head in the sand and remain an individual who works alone. It would be so easy to just quit on all this "teaming business."

However, the workplace is changing. The demands of the workplace today mandate that employees possess a highly honed set of team skills. Charles Darwin wrote, "It is not the strongest of the species that survive, nor the most intelligent, but the one most responsive to change." *The Team Developer* can help you in your journey of change. Carefully constructing a Personal Development Plan can help, too. So can reading this Guidebook. The rest is up to you. We suggest that you try to work through any real or perceived deficiencies. It is better to address your shortcomings now rather than later while on the job. Addressing them now may mean a lower course grade; addressing them on the job may translate into a denied promotion, lower merit increase, or termination.

People come in all shapes and sizes and colors. Physical characteristics that you can see are irrelevant. Psychological characteristics that you cannot see are paramount (e.g., values, motives, emotions, attitudes, personalities, work style preferences). Being an effective team member requires that you interact with all types of "characters." Although you may prefer that some students *not* be on your team, consider it an opportunity that they are there. In the

43

workplace, most times you cannot choose your co-workers or customers. The team skills you develop now will help you immensely later in life. The feedback you obtain via *The Team Developer* will enable you to understand your strengths and enhance your weak areas.

Figure 8.1 The team-building challenge

For most of us, the "Golden Rule" was an interpersonal principle that was passed on generation after generation. We were taught to treat others the way we wanted to be treated. However, the Golden Rule makes a huge assumption — that others are like us. The diverse work force of today has made the Golden Rule obsolete. Nowadays, a new principle is in order. The so-called "Platinum Rule" asserts we should treat others the way *they* want to be treated, not necessarily the way *we* want to be treated (Alessandra, O'Connor, & Alessandra, 1990). As you develop your team skills, this new rule is critical to remember. As you interpret the feedback you receive from *The Team Developer*, this rule becomes even more salient. Your success as a team member depends on interacting with others the way *they* desire, not necessarily in the manner in which *you* may be comfortable.

Your success as a future professional will depend on developing a comprehensive set of team skills needed in tomorrow's dynamic workplace. And in the words of Guillaume Appollinaire, your willingness to learn and evolve will permit you to fly off the edge of the cliff (and beyond!). Good luck on your personal journey.

REFERENCES

Alessandra, T., O'Connor, M. M., & Alessandra, J. (1990). *People smart: Powerful techniques for turning every encounter into a mutual win.* La Jolla, CA: Keynote Publishing.

Bergmann, T. J., & De Meuse, K. P. (1996). Diagnosing whether an organization is truly ready to empower work teams: A case study. *Human Resource Planning, 19(1),* 38-47.

Campion, M. A., Medsker, G. J., & Higgs, A. C. (1993). Relations between work group characteristics and effectiveness: Implications for designing effective work groups. *Personnel Psychology, 46,* pp. 823-850.

Davis, K. (1977). *Human behavior at work.* New York: McGraw-Hill.

De Meuse, K. P., & Erffmeyer, R. C. (1994). The relative importance of verbal and nonverbal communication in a sales situation: An exploratory study. *Journal of Marketing Management, 4,* 11-17.

Dumaine, B. (1994). The trouble with teams. *Fortune,* September 5, pp. 86-92.

Fisher, K. (1993). *Leading self-directed work teams: A guide to developing new team leadership skills.* New York: McGraw-Hill.

Hammer, M., & Champy, J. (1993). *Reengineering the corporation: A manifesto for business revolution.* New York: Harper Business.

Hargie, O. (1986). *A handbook of communication skills.* New York: New York University.

Harrington-Mackin, D. (1994). *The team building tool kit: Tips, tactics, and rules for effective workplace teams.* New York: AMACOM.

Joinson, C. (1999). Teams at work. *HR Magazine,* May, pp. 30-36.

Kayser, T. A. (1995). *Mining group gold: How to cash in on the collaborative brain power of a group (2nd ed.).* New York: McGraw-Hill.

Larson, C. E., & LaFasto, F. M. J. (1989). *TeamWork: What must go right / What can go wrong.* Newbury Park, CA: Sage.

Lawler, E. E., III. (1994). Total quality management and employee involvement: Are they compatible? *Academy of Management Executive, 8(1),* 68-76.

Lawler, E. E., III., & Cohen, S. G. (1992). Designing pay systems for teams. *ACA Journal,* Autumn, pp. 6-18.

Luft, J., & Ingham, H. (1963). *Group processes: An introduction to group dynamics.* Mayfield Publishing.

Manz, C. C., & Sims, H. P., Jr. (1993). *Business without bosses: How self-managing teams are building high-performing companies.* New York: John Wiley & Sons.

McGourty, J., Tarshis, L.A., & Dominick, P. (1996). Managing innovation: Lessons from world class organizations, *International Journal of Technology Management, 3-4.*

Mohrman, S. A., & Mohrman, A. M., Jr. (1997). *Designing and leading team-based organizations: A workbook for organizational self-design.* San Francisco: Jossey-Bass.

Nirenberg, J. S. (1989). *How to sell your ideas.* New York: McGraw-Hill.

Schein, E. (1988). *Process consulting: Its role in organization development (Vol. 1).* Reading, MA: Addison-Wesley.

Shaw, D. G., & Schneier, C. E. (1995). Team measurement and rewards: How some companies are getting it right. *Human Resource Planning, 18(3),* 34-49.

REFERENCES

Sheridan, J. H. (1990). America's best plants. *Industry Week*, October, pp. 27-44.

    The team-based organization. (1994). Chicago: A. T. Kearney.

Sundstrom, E., De Meuse, K. P., & Futrell, D. (1990). Work teams: Applications and effectiveness. *American Psychologist, 45*, pp. 120-133.

Thomas, K. W., & Kilman, R. H. (1974). *The Thomas-Kilman conflict mode instrument.* Tuxedo, NY: Xicom.

Thompson, L. (2000). *Making the team: A guide for managers.* Upper Saddle River, New Jersey: Prentice Hall.

Tuckman, B. W. (1965). Developmental sequence in small groups. *Psychological Bulletin, 63,* 384-399.

Tuckman, B. W., & Jensen, M. A. C. (1977). Stages of small-group development revisited. *Group and Organization Studies, 2*, pp. 419-427.

Wellins, R. S., Byham, W. C., & Wilson, J. M. (1991). *Empowered teams: Creating self-directed work groups that improve quality, productivity, and participation.* San Francisco: Jossey-Bass.

Wesner, M., & Egan, C. (1990, September). Self-managed teams in operator services. Paper presented at International Conference on Self-Managed Work Teams, Denton, TX.

ABOUT THE AUTHORS

Jack McGourty, Ph.D. is an Associate Dean for The Fu Foundation School of Engineering and Applied Science at Columbia University. He is also the Director of Assessment for the National Science Foundation's Gateway Engineering Education Coalition led by Drexel University. His main responsibilities include the institutionalization of assessment processes in all member institutions including: Columbia University, Cooper Union, Drexel University, New Jersey Institute of Technology, Polytechnic University, Ohio State University, and University of South Carolina. Dr. McGourty holds a Visiting Professorship at Drexel University and is a senior researcher and editor of a technology management newsletter for Stevens Institute of Technology, where he received his Ph.D. in Applied Psychology. His research interests focus on assessment processes as enablers for student learning, educational reform, and organizational innovation. Dr. McGourty is an active member in the American Society for Engineering Education, American Association for Higher Education, and the American Psychological Association. He has published several articles and book chapters on assessment and topics in education.

Kenneth P. De Meuse, Ph.D. is Professor of Management at the University of Wisconsin - Eau Claire. Previously, he was on the faculties of Iowa State University and the University of Nebraska. For the past decade, Dr. De Meuse has been investigating the "human side" of corporate restructuring and downsizing. More than 100 universities and 150 corporations have contacted him regarding his research work in this area. He has appeared on Cable News Network, Associated Press Radio, and National Public Radio and has been featured in *The Wall Street Journal, Industry Week, Across the Board, Business Week, U.S. News & World Report*, and *USA Today* for his expertise on the impact corporate transitions have on employees. He has published articles in such journals as *Human Resource Management, Human Resource Planning, Group and Organizational Management, Human Relations*, and *American Psychologist*. Dr. De Meuse received his doctorate in Industrial/Organizational Psychology from the College of Business at The University of Tennessee.

Appendices

1. Introduction to *The Team Developer* Report

2. Sample Report

3. Student Disk Instructions

INTRODUCTION TO THE TEAM REPORT

First of all, congratulations! You are about to receive feedback that will help you to become a more effective team member. Your report will summarize information provided by you and your fellow team members in response to a recent *Team Developer* survey. In addition, for the first time, you can compare other team members' ratings of you from this administration of the survey and from the last administration. Most of your feedback is being provided according to four team effectiveness dimensions: Collaboration, Communication, Decision Making, and Self Management. Each dimension is defined in greater detail later in the report.

HOW TO USE YOUR REPORT

Receiving feedback is always helpful, but sometimes the process can be difficult. In order to get the most out of the information in this report, consider the following:

- As stated in the introduction, the purpose of this report is to assist in your development. Therefore, focus on how this information can help you improve and solve problems.

- Your feedback is essentially a "snapshot" of how you and others perceive your behavior. Whether this picture changes is largely up to you.

- You may be surprised by how others have assessed your performance as a team member. Even if you do not agree, accept the feedback as their perceptions and try to acknowledge that who you appear to be is at least part of who you are.

Use this report as a tool to help you further explore your performance as a team member. As you review the report, compare your self-ratings with how the other team members rated you. Also, look at any changes in how others rated you from the last administration and follow up on the report's feedback by seeking clarification from others. Finally, use the feedback and the development suggestions to help define specific action steps you can take to improve your effectiveness.

INTERPRETING THE RESULTS

Rating Scale:

All numbers that appear in this report are presented in terms of the five-point scale that appeared on *The Team Developer*. An interpretation of the scale points is shown below.

```
Above 4.0  = Strength

 3.0 - 4.0  = Adequate
Effectiveness

Below 3.0  = Development Area
```

The scale interpretation (strength, adequate effectiveness, or development area) applies to ratings for dimensions and individual items.

Rounding:

Because results are based upon all available data and are rounded to one decimal place, the average rating shown for a single dimension may not always equal the result of averaging the numbers shown for the items within that dimension.

Dimension Definitions:

As mentioned in the Introduction, the items that make up *The Team Developer* are grouped into four dimensions. Each dimension is defined below.

Collaboration

Demonstrating a commitment to the team's overall purpose, helping team members to identify mutual objectives, working cooperatively and constructively with others both inside and outside the team, actively participating in team activities, showing support and encouragement for fellow team members.

Communication

Helping to sustain an environment where people feel free to speak candidly, articulating ideas clearly and concisely, listening and demonstrating an understanding of others' perspectives.

Decision Making

Gathering information and weighing alternatives when addressing an issue, working with the team toward resolution, promoting innovative thinking, ensuring that a rationale forms the basis for the decisions made.

Self Management

Utilizing appropriate styles, methods and procedures to direct individuals and the team toward goal achievement; modeling and modifying behavior as required to achieve results while being sensitive to individual and group processes.

DEVELOPMENTAL SUGGESTIONS

This section provides several recommendations for improving your performance in each team effectiveness dimension. You should especially focus your attention on the lowest rated dimensions in Section 3, your Dimension Overview. As you read through the suggestions, highlight or place a checkmark beside those that seem most relevant. Remember, these are only suggestions to guide your developmental efforts. You also should consider getting additional suggestions from team members, friends, and your professor.

COLLABORATION

Managing Conflict:

- Look at conflict as a difference of ideas and opinions rather than as a personality issue.

- State your points clearly and concisely. Avoid lecturing to convince others that you are right.

- Get feedback from peers and colleagues about your effectiveness in handling conflict in group situations.

- Allow others to vent their frustrations. This will help people, including you, get down to problem-solving. Try to talk about your frustrations rather than show them.

- Use a neutral third party (such as a peer, a human resources representative, or a group process observer) to help work through problems.

- Paraphrase the positions held by others to ensure that the conflict is not just a misunderstanding and to show that you understand their perspectives.

- Look for underlying causes that may be the basis for conflict. If you think there may be other causes, quietly ask someone questions like, "Is something else bothering you?" or "Have we clarified all the issues?"

- If tensions are getting too high, ask for a break so that people can cool off and the discussion can remain level-headed.

- Try to identify objectives and perspectives you have in common with the other party. Build upon the things you have in common to come up with potential win/win solutions.

Recommended Readings:

Collaborating: Finding Common Ground for Multi-Party Problems by Barbara Gray. San Francisco: Jossey-Bass, 1989.

Conflict in Organizations: Practical Solutions Any Manager Can Use by Steve Turner & Frank Weed. Englewood Cliffs, NJ: Prentice-Hall, 1983.

Managing Conflict: Interpersonal Dialogue and Third Party Roles (2nd ed.) by Richard E. Walton. Reading, MA: Addison-Wesley, 1987.

The Wisdom of Teams: Creating the High Performance Organization by Jon R. Katzenbach & Douglas K. Smith. Harperbusiness, 1994.

Results-Based Leadership by David Ulrich, Jack Zenger, & Norm Smallwood. Harvard Business School Press, 1999.

Power and Organization Development: Mobilizing Power to Implement Change (Addison-Wesley OD Series) by Larry E. Greiner & Virginia E. Schein. Addison-Wesley, 1988.

Mastering the Art of Creative Collaboration (Businessweek Books) by Robert Hargrove. McGraw-Hill, 1998.

Developing Management Skills: Managing Conflict by David A. Whetten & Kim S. Cameron. Addison-Wesley Publishing Co., 1993.

Resolving Conflict at Work: A Complete Guide for Everyone on the Job by Kenneth Cloke & Joan Goldsmith. San Francisco: Jossey-Bass, 2000.

The Handbook of Conflict Resolution: Theory and Practice by Morton Deutsch & Peter T. Coleman. San Francisco: Jossey-Bass, 2000.

Creating a Team Environment:

- Welcome opportunities to collaborate with others on tasks or projects.
- When helping to plan team objectives, try to determine and incorporate people's personal goals.
- Compliment others when you think they have done a good job.
- Look for ways to involve other team members during discussions. Open ended-questions and reflective listening can be helpful.
- Actively seek others' input on and opinions of your work.
- Become familiar with the roles and responsibilities of fellow team members.
- Avoid pre-judging others' ideas and suggestions.
- Learn about others by asking them about their interests. Similarly, share information about yourself that is beyond the scope of the team's project.
- Seek small opportunities to build acquaintances with team members (e.g., coffee, lunch, planned social activities).
- Offer to help another team member who seems to have a lot of work or is struggling with a difficult problem.

Recommended Readings:

Team Players and Teamwork: The New Competitive Business Strategy by Glenn Parker. San Francisco: Jossey-Bass, 1990.

Developing Superior Work Teams: Building Quality and the Competitive Edge by D.C. Kinlaw. San Diego: University Associates Inc., 1991.

Team-Based Organizations: Developing a Successful Team Environment by J. Schonk. Homewood, IL: Business One Irwin, 1992.

The Team Handbook by P.R. Scholtes. Wisconsin: Joiner Associates, 1988.

The Different Drum: Community-Making & Peace by M. Scott Peck. New York: Touchstone Books, 1988.

Rewarding Teams: Lessons from the Trenches by Glenn M. Parker, Jerry McAdams, & David Zielinski. San Francisco: Jossey-Bass, 2000.

For Team Members Only: Making Your Workplace Team Productive and Hassle-free by Charles C. Manz, Christopher P. Neck, James Mancuso, & Karen P. Manz. New York: Amacom, 1997.

Tips for Teams: A Ready Reference for Solving Common Team Problems (McGraw-Hill Training Series) by Kimball Fisher, Steven Rayner, & William Belgard. New York: McGraw-Hill, 1995.

Designing and Leading Team-Based Organizations: A Workbook for Organizational Self-Design by Susan Albers Mohrman & Allan M. Mohrman, Jr. San Francisco: Jossey-Bass, 1997.

The Team Handbook (2nd edition) by Peter R. Scholtes, Brian L. Joiner, & Barbara J. Streibel. Madison, WI: Oriel Inc, 1996.

Commitment to Team Goals:

- Make certain that you are in agreement with others about the team's overall purpose and major objectives.

- Try to determine how your personal goals relate to the team's goals. Make a list of the ways they are consistent with one another. Make a separate list of any ways they are incompatible.

- Learn about other members' personal goals. Look for ways to integrate them with your own. Do the same with team goals.

- Think about what other sorts of changes or activities would help you feel more enthusiastic about your work. Determine which ones are directly under your control and which require the assistance of others. Develop a plan to make some of these changes and activities happen.

- Let others know that you are interested in challenging projects and personal growth.

- Demonstrate initiative by suggesting ways to enhance or modify team objectives and going beyond what is expected on projects/tasks.

Recommended Readings:

Thriving on Chaos: Handbook for Management by Tom Peters. New York: Alfred A. Knopf, 1988.

Principle Centered Leadership by Stephen R. Covey. New York: Simon & Schuster, 1990.

The Seven Habits of Highly Effective People by Stephen R. Covey. New York: Simon & Schuster, 1989.

Goal Setting: A Motivational Technique That Works! by Edwin A. Locke & Gary P. Latham. Englewood Cliffs, NJ: Prentice-Hall, 1984.

A Passion for Excellence: The Leadership Difference by Tom Peters & Nancy K. Austin. New York: Warner Books, 1989.

The Power Principle: Influence with Honor by Blaine Lee & Stephen R. Covey. New York: Fireside, 1998.

The Nature of Leadership by Stephen R. Covey, A. Roger Merrill & Dewitt Jones. Provo, UT: Covey Leadership Center, 1998.

The Seven Habits of Highly Effective People: Powerful Lessons in Personal Change by Stephen R. Covey. New York: Franklin Covey Co., 1990.

Make Success Measurable!: A Mindbook-Workbook for Setting Goals and Taking Action by Douglas K. Smith. New York: John Wiley & Sons, 1999.

Coaching and Support:

- Acknowledge people's feelings as having a significant impact on the way they perform.
- Compliment others when you think they have done a good job.
- Provide feedback and constructive criticism when you feel someone has not done well, but avoid being overly judgmental in your day-to-day interactions.
- Ask others directly how things are going for them.
- Communicate in a non-threatening way that you are available and willing to provide feedback and expertise.
- Try to determine the kinds of resources that others may need to complete a task. Do what you can to help them obtain what is needed.
- Pass along books, articles, and other developmental suggestions to others that you think may be helpful.
- Encourage people to focus on constant improvement by acknowledging their efforts to correct mistakes or enhance their skills.

Recommended Readings:

People Skills by Robert Bolton. New York: Touchstone Books, 1986.

A Passion for Excellence by T. Peters & N. Austin. New York: Random House, 1985.

Bringing Out the Best In People by Alan Loy McGinnis. Minneapolis, MN: Augsburg Publishing House, 1985.

Executive Talent: How to Identify & Develop the Best by Tom Potts & Arnold Sykes. Homewood, IL: Business One Irwin, 1993.

101 Ways to Improve Your Communication Skills Instantly by Jo Condrill & Bennie Bough. Palmdale, CA: Goalminds, 1999.

Bringing Out The Best in People: How To Apply the Astonishing Power of Positive Reinforcement (2nd edition) by Aubrey C. Daniels. New York: McGraw-Hill, 1999.

Performance Management: Improving Quality and Productivity Through Positive Reinforcement (3rd edition) by Aubrey C. Daniels. Performance Management Publishing, 1989.

Leadership Development: Building Executive Talent by American Productivity and Quality Center, Houston, TX: 1999.

Executive Talent: Developing and Keeping the Best People by Eli Ginzberg. Transaction Publishing, 1995.

COMMUNICATION

Providing and Accepting Feedback

Providing Feedback:

- When giving feedback, focus your comments on the specific issues and behaviors, not the person. At no point should the person's self-worth be questioned or challenged.
- Avoid describing people's performance in terms of general traits. Instead, give specific, behavioral examples to make your point clear.
- Try to keep feedback immediate. Avoid long delays between the time something happens and your response.
- Let team members know how their behavior is affecting the behavior of others and the team's ability to accomplish its objectives.
- Provide positive feedback as well as negative feedback. Try to provide positive information first.
- When possible, let people know up-front (before they begin a project or task) what's important and what they will be evaluated on.

- Involve the person receiving feedback in identifying problems and developing solutions. Plan on encouraging the person receiving feedback to speak at least as much as you do. Ask open-ended questions such as, "What do you think went well?" or "What do you think should be done differently?"

Accepting Feedback:

- Focus on how feedback can help you deal with particular problems. Try not to perceive it as personal criticism.
- To reduce defensiveness when receiving feedback, stay focused on the issues. Ask yourself: "Do I understand what is being said?" "Can I do something about this issue?" "What would happen if I acted on this feedback?" "What additional information would be helpful?"
- Restate to people what you think they have said to demonstrate or confirm understanding.
- Make it a point to regularly solicit informal feedback from your team members and supervisor.
- Even if you disagree with feedback, accept it as reality for the person giving it.

Recommended Readings:

The One-Minute Manager by Kenneth Blanchard & Spencer Johnson. New York: Berkeley Publishing Group, 1987.

A Handbook of Communication Skills by O. Hargie. New York: New York University, 1986.

Goal-Setting: A Motivational Technique that Works! by Edwin A. Locke & Gary P. Latham. Englewood Cliffs, NJ: Prentice-Hall,1984.

Leadership and the One-Minute Manager: Increasing Effectiveness Through Situational Leadership by Patricia Zigarmi, Drea Zigarmi & Kenneth H. Blanchard. William Morrow & Co., 1985.

Effective Listening Skills (Business Skills Express) by Dennis M. Kratz & Abby Robinson Kratz. Burr Ridge, IL: Irwin Professional Publishing, 1995.

Influencing Others:

- Try modeling the techniques and behaviors of those you consider to be highly influential.
- Prepare for meetings in advance. Try to determine ahead of time where you think you can make important contributions.
- Look for opportunities that give you a chance to lead a group and influence others.
- Ask peers/colleagues for feedback on how persuasive and influential you are. Ask them how you can become more so.
- Make a commitment to speak your concerns during meetings and use clear, concise language to communicate your position. Keep a log of issues on which you do and don't mention your concerns. Include in the log the reasons you think you did or didn't fully express your position.
- Be one of the first people to offer ideas in meetings.
- View yourself as someone who has something valuable to contribute to others. Think of yourself as a leader.
- Be prepared with data and facts to help make your points.
- When negotiating, try to state your position in positive terms. For example, talk in terms of what you will do rather than what you won't do.

Recommended Readings:

Power & Influence: Beyond Formal Authority by John P. Kotter. New York: The Free Press, 1986.

How to Sell Your Ideas by Jesse S. Nirenberg. New York: McGraw-Hill, 1989.

Influence Without Authority by Allan C. Cohen & David L. Bradford. New York: John Wiley & Sons, 1990.

Getting to Yes: Negotiating Agreement Without Giving In by Roger Fisher & William Ury. New York: Penguin Books, 1981.

Getting Together: Building Relationships as We Negotiate by Roger Fisher & Scott Brown. New York: Penguin Books, 1988.

You've Got to Be Believed to Be Heard by Bert Decker. New York: St. Martin's Press, 1992

How to Create High Impact Business Presentations by Joyce Kupesh & Pat R. Graves. Lincolnwood, IL: NTC Business Books, 1993.

The Leadership Factor by John P. Kotter. New York: Free Press, 1988.

Managing with Power: Politics and Influence in Organizations by Jeffrey Pfeffer. Boston, MA: Harvard Business School Publishing, 1996.

Power Up: Transforming Organizations Through Shared Leadership by David L. Bradford & Allan R. Cohen. New York: John Wiley & Sons, 1998.

Managing for Excellence: The Guide to Developing High Performance in Contemporary Organizations (Wiley Management Classics) by David L. Bradford & Allan R. Cohen. New York: John Wiley & Sons, 1997.

Getting Past No: Negotiating Your Way From Confrontation to Cooperation by William Ury. New York: Bantam Doubleday Dell Publishing, 1993 (revised edition).

Beyond Machiavelli: Tools for Coping with Conflict by Roger Fisher, Elizabeth Kopelman, & Andrea Kupfer Schneider. New York: Penguin USA, 1996.

DECISION MAKING

Planning and Analysis:

- Prioritize the important decisions you have to make during a particular time period. Work on them in order.
- Don't try to solve large or complex problems all at once. Plan to address them over an extended period of time.
- Gather as much information as possible, but avoid getting bogged down in details.
- Seek input from those closest to a problem and those most likely to be affected by the way it is solved. Determine in advance who those individuals or groups are.
- From time to time, list all the possible solutions in order to help you identify additional alternatives.
- Try to rephrase a problem in different words to help uncover alternatives.
- Take the time to double check the facts, figures, and data upon which you will make a decision.

Recommended Readings:

Brain Power: Learn to Improve Your Thinking Skills by Karl Albrecht. Englewood Cliffs, NJ: Prentice Hall, 1987.

The Mind of the Strategist: Business Planning for Competitive Advantage by Kenichi Ohmae. New York: McGraw-Hill, 1982.

Strategic Management & Organizational Decision Making by Alan Walter Steiss. Lexington, MA: Lexington Books, 1985.

Effective Business Decision Making by William F. O'Dell. Lincolnwood, IL: NTC Business Books, 1991.

The Agile Manager's Guide to Making Effective Decisions (The Agile Manager Series) by David F. Folino. Bristol, VT: Velocity
> Publishing, 1998.

Great Critical Thinking Puzzles by Michael A. Dispezio. New York: Sterling Publications, 1997.

Innovation and Risk Taking:

- Avoid making decisions without thinking of more than one alternative. To help you generate a list of possibilities, take an occasional break. When you resume working, begin by redefining the problem and try approaching it from a different perspective.
- Draw pictures of problems instead of writing them down.
- Read material outside of your main areas of expertise or normal interests. This may help you to develop new perspectives on problems.
- When evaluating alternatives, ask yourself (and others) "Why not?" instead of "Why?"
- Talk with others inside and outside your team to see how they have solved similar problems.
- Try not to take yourself too seriously when working on a problem.
- Think of yourself as innovative and creative.
- Try to approach other team members with possible solutions rather than just presenting them with the problem.

Recommended Readings:

A Whack on the Side of the Head: How to Unlock Your Mind for Innovation by Roger von Oech. New York:
> Warner Books, 1983.

If It Ain't Broke...Break It by Robert J. Kriegel & Louis Patler. New York: Warner Books, 1991.

The Creative Edge by William C. Miller. Reading, MA: Addison-Wesley, 1990.

40 Years, 20 Million Ideas: The Toyota Suggestion System by Yugo Yamada. Cambridge, MA: Productivity
> Press, 1991.

A Whack on the Side of the Head: How You Can Be More Creative by Roger Von Oech. New York: Warner Books,
1998.

What a Great Idea!: The Key Steps Creative People Take by Charles "Chic" Thompson. New York: Harper Perennial
Library, 1992.

Don't Compete...Tilt the Field by Louis Patler. Oxford, UK: Capstone Ltd., 1999.

Breakthrough Thinking: The Seven Principles of Creative Problem Solving by Gerald Nadler & Shozo Hibino.
> Rocklin, CA: Prima Publishing, 1998.

Judgment/Using Facts:

- To avoid making decisions too quickly, first assess whether an immediate response is absolutely necessary. Ask yourself, "What is the worst thing that could happen if I waited to make this decision?"
- If you tend to delay making decisions, set time limits for deciding. Push yourself to meet the deadline. You may want to begin by practicing with some relatively low-risk decisions. After you have built up confidence, use the approach for more important decisions.
- Accept the fact that there is a certain amount of risk in making any decision.
- Be willing to change your position if new information is presented later.
- Try to approach other team members with possible solutions rather than simply presenting them with the problem. Try to determine whether they agree or disagree with the solutions you've proposed.

- View problems in terms of how they will impact the people involved.
- Accept the fact that you will seldom have all the information you would like to have.

Recommended Readings:

Making Judgments, Choices & Decisions in Business: Effective Management Through Self-Knowledge by Warren J. Keegan. New York: John Wiley & Sons, 1984.

Managerial Decision Making by George P. Haber. Glenview, IL: Scott Foresman & Company, 1980.

Whatever it Takes: Decision Makers at Work by Morgan W. McCall Jr. & Robert E. Kaplan. Englewood Cliffs, NJ: 1985.

Judgment In Managerial Decision Making (2nd ed.) by Max H. Boyerman. New York: John Wiley & Sons, 1990.

Judgment in Managerial Decision Making (4th edition) by Max H. Bazerman. New York: John Wiley & Sons, 1997.

The Psychology of Judgment and Decision Making (McGraw-Hill Series in Social Psychology) by Scott Plous. New York: McGraw-Hill College Division, 1993.

Making Choices: A Recasting of Decision Theory by Frederic Schick. Cambridge, UK: Cambridge University Press, 1997.

Whatever It Takes: The Realities of Managerial Decision Making by Robert E. Kaplan & Morgan W. McCall. Englewood Cliffs, NJ: Prentice Hall College Division, 1992.

SELF-MANAGEMENT

Establishing Direction and Standards:

- Establish goals, priorities, and timelines before beginning a new project.
- Communicate your plan to others and be open to suggestions.
- Make certain that new team members understand their individual roles and the team's overall purpose.
- Let people know up-front what you expect from them.
- Identify things you like and don't like about a particular change. Try to determine why you feel this way.
- View your and others' resistance to change as a problem to solve, not as a character flaw.

Recommended Readings:

The Purpose Driven Organization: Unleashing the Power of Direction and Commitment by Perry Pascarella & Mark A. Frohman. San Francisco: Jossey-Bass, 1989.

The Change Masters by Rosabeth Moss Kanter. New York: Simon & Schuster, 1983.

Thriving on Chaos: Handbook for Management Revolution by Tom Peters. New York: Alfred A. Knopf, 1988.

Goal Setting: A Motivational Technique That Works! by Edwin A. Locke & Gary P. Latham. Englewood Cliffs, NJ: Prentice-Hall, 1984.

When Giants Learn to Dance by Rosabeth Moss Kanter. New York: Touchstone Books, 1990.

The Challenge of Organizational Change: How Companies Experience It and Leaders Guide It by Rosabeth Moss Kanter, Barry A. Stein, & Todd D. Jick. New York: Free Press, 1992.

A Passion for Excellence: The Leadership Difference by Tom Peters, Nancy K. Austin, & Thomas J. Peters. New York: Warner Books, 1989.

Goal Setting 101: How to Set and Achieve a Goal! by Gary Ryan Blair. Tarpon Springs, FL: The GoalsGuy, 2000.

Empowering Others:
- Share your success as an individual with the team.
- Involve other team members in every aspect of the strategic process. Obtain their input on the team's vision, strategy, objectives, and tactics.
- Continually review the key strengths and weaknesses of each person on your team. Look for ways to best utilize their strengths.
- Seek the input and opinions of others when defining problems, developing alternatives, and making decisions.
- Make a list of the activities for which you are responsible. Determine which ones are those only you can perform, those that can be shared with others and those that can be delegated. From time to time, review the extent to which you have shared or delegated the appropriate activities.
- Make a conscious effort to entrust others with the appropriate authority and responsibility. Let people know when you believe in their ability to get the job done.

Recommended Readings:

No-Nonsense Delegation by D.D. McConkey. New York: AMACOM, 1979.

"Managers and Leaders: Are They Different?" by Abraham Zaleznik. *Harvard Business Review,* March/April 1992.

The Empowered Manager: Positive Political Skills at Work by Peter Block. San Francisco: Jossey-Bass, 1987.

High Involvement Management by Edward E. Lawler III. San Francisco: Jossey-Bass, 1987.

The 3 Keys to Empowerment: Release the Power Within People for Astonishing Results by Kenneth H. Blanchard, John P. Carlos, & Alan Randolph. San Francisco, CA: Berrett-Koehler Publishers, 1999.

Empowering Employees Through Delegation by Robert B. Nelson. Burr Ridge, IL: Irwin Professional Publishing, 1993.

Personal Integrity and Respect
- Make a list of some of your most important convictions. Develop another list describing how those convictions have influenced decisions you have made. Develop the same kind of list for decisions you will soon be making.
- When taking a contrary position to others, start by saying that your view may be unpopular and then explain why you feel it is an important perspective.
- Develop a habit of arriving 10-15 minutes early for meetings and appointments.
- Help to ensure that you follow through on commitments by establishing deadlines and noting them on your calendar. Learn to say "no" if there is a good chance you won't be able to follow through on a request.
- Lead by example.
- Be prompt in responding to people's phone calls, notes, and other requests for information.

Recommended Readings:

Principle Centered Leadership by Stephen R. Covey. New York: Simon & Schuster, 1990.

Right on Time! The Complete Guide for Time-Pressured Managers by Lester R. Bittel. New York: McGraw-Hill, 1991.

The Seven Habits of Highly Effective People by Stephen R. Covey. New York: Simon & Schuster, 1989.

Off the Track: Why & How Successful Managers Get Derailed by Morgan W. McCall Jr. & Michael M. Lombardo. Westport, CT: Greenwood Press, 1985.

Managers Talk Ethics: Making Tough Choices in a Competitive Business World by Barbara Ley Toffler. New York: John Wiley & Sons, 1991.

On Becoming a Leader by Warren Bennis. Reading, MA: Addison-Wesley, 1989.

The Power Principle: Influence with Honor by Blaine Lee & Stephen R. Covey. New York: Fireside, 1998.

The Seven Habits of Highly Effective People: Powerful Lessons in Personal Change by Stephen R. Covey. New York: Fireside, 1990.

Learning to Lead: A Workbook on Becoming a Leader by Warren Bennis & Joan Goldsmith. Cambridge, MA: Perseus Publishing, 1997.

Leaders: Strategies for Taking Charge by Warren Bennis & Burt Nanus. New York: HarperBusiness, 1997.

First Things First: To Live, to Love, to Learn, to Leave a Legacy by Stephen R. Covey, A. Roger Merrill & Rebecca R. Merrill. New York: Fireside, 1996.

First Things First Every Day: Because Where You're Headed is More Important Than How Fast You're Going by Stephen R. Covey. New York: Fireside, 1997.

Technical Learning:

- Ask your team members and co-workers for feedback on your two or three strongest and weakest technical areas.
- Volunteer for technically challenging projects and ask for assignments outside of your own functional area.
- Join professional associations in your field and attend seminars and conferences to help you build skills.
- Make sure your team members know about your areas of technical expertise and tell them you are available to help them learn and to work on projects. Similarly, ask others to help you in their areas of expertise.
- Identify potential mentors outside of your management chain who can help you learn and develop.
- Stay on top of changes in technology by reading professional publications, talking with others in your field, and attending workshops. Look for ways you and your team can apply these innovations.
- Regularly look over library listings for new books and other materials related to your field. Ask library staff to contact you when they receive something related to your line of work or area of interest. In addition, let other team members know when you come across materials they might find interesting or useful.
- Take advantage of electronic networks and information services that allow you to receive information about specific topics and communicate with others who share your interests.

Recommended Readings:

No reading materials are listed for this section because of the specialized nature of many occupational and technical areas. For detailed information about your area of expertise, contact your library, professional/trade associations, local universities, and the training department in your organization.

DEVELOPMENT PLAN

Now, it's time to decide how you are going to act on the information provided in this report. Use the planning sheet on page 61 to address the development areas identified in your feedback. If necessary, make copies of the form.

To use the planning sheet, describe the development area you want to focus on in the first column. Rather than working on an entire dimension all at once, try to select development areas related to specific survey items. In the second column, list the action steps you will take to improve your performance. You may want to look back to the development suggestions you checked or highlighted as you do this. Also, obtain additional input from team members, friends, or your professor to help you prepare more effective action steps. For example, your professor, who is familiar with your report, may be able to involve you in projects that will aid in your development or direct you toward additional resources. Finally, record your target completion dates in the third column.

You will have more success in improving your performance if your action steps define *precisely* what you are going to do. You want to avoid vague, general action steps that are difficult to measure. With clear action steps, you can determine whether you've met, exceeded, or fallen below your objectives. It also helps to have more than one action step for a particular development area. Examples of vague and specific action steps are shown below to guide you as you write your own development plan.

Development Area	Action Steps	Completion Date
Persuading others to adopt my views.	Vague: *Present better. Think out arguments in advance of each meeting.*	*5/14*
	Specific: *1) Prepare in advance a fact-based argument to support my position on XYZ issue. Seek time to present my position at our next meeting.*	*5/14*
	2) After the meeting, seek feedback from two other people on the impact of my argument.	*5/18*
	3) Read the book, How to Sell Your Ideas.	*5/31*
Generating alternative solutions.	Vague: *Be more creative by thinking about things from different perspectives and talking to people from other backgrounds.*	*6/30*
	Specific: *1) Prior to making a decision on XYZ, seek the input of at least three people whose backgrounds are different from one another and from my own.*	*6/15*
	2) During the next month, keep a notebook handy to write down new ideas. Review my ideas once a week with a view toward how they will help our team solve problems.	*6/30*

	TEAM DEVELOPER	
	DEVELOPMENT PLAN	
Development Area	Action Steps	Completion Date

The Team Developer:
An Assessment and
Skill Building Program

Student Feedback Report
(Sample)

Team Developer Results

Results for:

Team:	Self	Team

COMMUNICATION **3.70** **3.78**

    Listens attentively to others without interrupting 4.00   4.25

    Conveys interest in what others are saying 3.00   4.25

    Provides others with constructive feedback 3.00   3.50

    Restates what has been said to show understanding 3.00   3.00

    Clarifies what others have said to ensure understanding 3.00   3.50

    Articulates ideas clearly and concisely 4.00   4.00

    Uses facts to get points across to others 5.00   3.50

    Persuades others to adopt a particular point of view 4.00   3.75

    Gives compelling reasons for ideas 4.00   4.25

    Wins support from others 4.00   3.75

DECISION MAKING 4.13   3.60

    Analyses problems from different points of view 4.00   3.50

    Anticipates problems and develops contingency plans 4.00   3.25

    Recognizes interrelationships among problems and issues 4.00   3.75

    Reviews solutions from opposing perspectives 4.00   3.75

    Applies logic in solving problems 5.00   3.75

    Challenges the way things are being done 5.00   3.50

    Solicits new ideas from others 4.00   3.50

    Generates new ideas 4.00   4.00

    Accepts change 4.00   4.00

    Suggests new approaches to solving problems 4.00   4.00

    Offers solutions based on facts rather than "gut feel" or intuition 4.00   3.50

    Discourages others from rushing to conclusions without facts 4.00   3.25

    Organizes information into meaningful categories 4.00   3.50

    Helps others to draw conclusions from the facts 3.00   3.25

    Brings in information from "outside" sources to help make decisions 5.00   3.50

COLLABORATION **3.90** **3.90**

    Acknowledges issues that the team needs to confront and resolve 4.00   4.00

    Encourages ideas and opinions even when they differ from his/her own 3.00   3.25

Team Developer Results

Results for:

	Self	Team
Team:		

COLLABORATION **3.90** **3.90**

Works toward solutions and compromises that are acceptable to all **4.00** 3.75
Helps reconcile differences of opinion **3.00** 3.25
Accepts criticism openly and non-defensively **3.00** 3.50

Shares credit for successes with others **4.00** 4.75
Cooperates with others **4.00** 4.75
Encourages participation from all involved **5.00** 4.00
Shares information with others **5.00** 4.25
Reinforces the contributions of others **4.00** 3.50

SELF-MANAGEMENT **4.33** **3.30**

Monitors progress to ensure that goals are being met **4.00** 3.25
Creates action plans and timetables for work session goals **4.00** 2.75
Defines task priorities for work sessions **4.00** 2.75
Ensures that goals are understood by all **5.00** 3.00
Puts top priority on getting results **5.00** 3.50

Stays focused on the task during meetings **3.00** 3.75
Uses meeting time efficiently **4.00** 3.25
Suggests ways to proceed during work sessions **5.00** 3.25
Clarifies roles and responsibilities of others **4.00** 2.75
Reviews progress throughout work sessions **4.00** 3.25

Solicits input from all members **5.00** 3.75
Encourages frequent polling among team members **5.00** 3.50
Summarizes the team's position on issues **4.00** 3.25
Involves others in decisions that affect them **5.00** 3.50
Encourages others to express their views even when they are contrary **4.00** 4.00

Team Developer
Instructions for using the Team Developer Student Disk

How to start and navigate through *The Team Developer* Assessment and Skill Building Program provided by your instructor?

Using *The Team Developer* Student Disk is quite easy. Hundreds of students have used the Team Developer software without any problems. All you need to do is follow a few simple steps once your instructor has given you the survey floppy disk. However, you can begin right away and practice on the sample student application located at www.wiley.com/college/mcgourty or you can follow these instructions for the disk supplied by your instructor.

Logging on to *The Team Developer*

The first step is to insert *The Team Developer* Student Disk (3.5 floppy) into drive A. Click on the Windows START button and then click on RUN. Where it states Open, type in A: WTD and then click on OK.

The Team Developer splash screen will appear followed by the LOGON screen (see Exhibit #1). What you will see is the Alphabet preceded by a small plus sign "+". Go to the letter of your last name and click on the + sign. (If you are running the SAMPLE application on the Web, click on any letter and select any name.) Once you click on the + sign next to the letter of your last name, one or more names will appear – all starting with the selected letter and in alphabetical order. You should locate your name and click on it. The LOGON button now becomes highlighted.

Exhibit #1

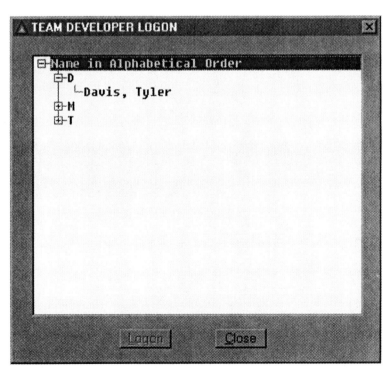

65

Once you click on the LOGON button, a screen will appear stating what team you belong to and your name (see Exhibit #2). If these are correct, you may proceed. (If either the name of your team or your name is incorrect, click on CANCEL and, then CLOSE. You will need to contact your instructor for further information.) Assuming that the team and your name is correct, you must enter a password and the last four digits of your social security number. The starting password for all students is "TEAM." (You will be able to change the password to your own personal one later in the program.) Click on OK once you have entered the password.

Exhibit #2

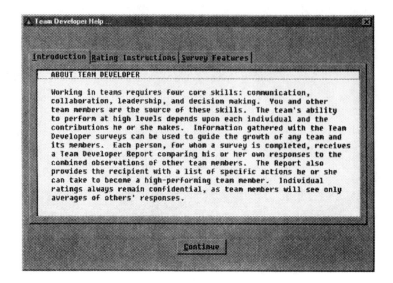

After entering the password, type in the last four digits of your social security number; for example, 1234. Again, click on the OK button. You will have now successfully logged on to *The Team Developer*. Upon entering the program, you will be provided with a short help screen (see Exhibit #3) with a choice of three categories of information. 1) Introduction gives you a quick synopsis of *The Team Developer*; 2) Rating Instructions provides some useful guidelines to think about while rating yourself and fellow team members; and 3) Survey Features includes comments from your instructor.

Exhibit #3

Rating Yourself and Fellow Team Members

Once you have reviewed the information provided in *The Team Developer* Help screen, you are ready to begin the survey process. Click CONTINUE to go to the survey portion of the program.

You will notice that as you begin the survey, a dialogue box appears in the middle of your screen (see Exhibit #4). This box contains a definition of the skills and behaviors on which you will be rating yourself and fellow team members over the next few survey items or questions. These definition dialogue boxes will appear on your screen to alert you when it is time for you to think about survey items and questions within a specific context. For example, when responding to survey items relating to communication, you should be thinking about situations in which you and fellow team members were discussing team topics or presenting results to each other. You can refer to these definitions anytime while answering specific survey questions. Just click on the DEFINITION button displayed in the upper left-hand corner of the screen.

Exhibit #4

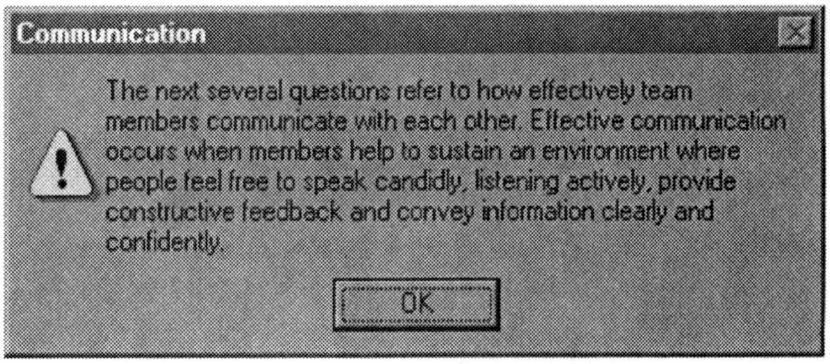

After you have read the skill or behavior definition in the dialogue box presented, click OK. You can now proceed to rate yourself and your fellow team members on the first question (see Exhibit #5). Typically, the first thing to do is think about the survey question in terms of how frequently you exhibit the behavior presented. For example, if the question states: "Listens attentively without interrupting," you will want to honestly evaluate how often you do indeed listen attentively to other team members. Or how often you find youself interrupting while others are speaking. Once you have reflected on the times you have worked together as a team, click on the number that best represents the frequency by which you exhibit this behavior. If you feel that you always, one hundred percent of the time listen to others attentively and never interrupt, click on the "5". However, if you frequently interrupt and never listen to anyone without starting to speak yourself, then you should rate yourself a "1".

Exhibit #5

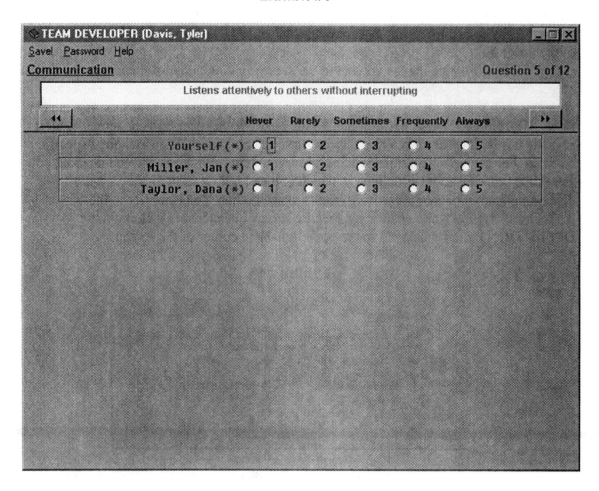

Once you have rated yourself on a specific survey question, go through the same reflective process for each team member. Consider how frequently each member exhibits the behavior in question, and rate them accordingly. After you have rated yourself and all other members, you can quickly review the ratings you have given as they are displayed immediately after each name (see Exhibit #6). If you are satisfied with your ratings, you can move to the next question. To do so, click on the ARROW located in the upper right hand of the screen. Remember you can always go back and make changes to your ratings.

You should continue the above process until you have completed all the survey questions for yourself and each of your fellow team members. Once you have completed all the questions, you must save your work before leaving the program. If you wish to stop and take a brake prior to completing all the questions, you also must save your work.